The Universe Next Door

A Journey Through 55 Alternative Realities, Parallel Worlds and Possible Futures

Books by New Scientist *include*

Does Anything Eat Wasps?
Why Don't Penguins' Feet Freeze?
How to Fossilise Your Hamster
Do Polar Bears Get Lonely?
How to Make a Tornado
Why Can't Elephants Jump?
Farmer Buckley's Exploding Trousers
Why Are Orangutans Orange?
Will We Ever Speak Dolphin?
Nothing
Question Everything
Chance
How Long is Now?

The Universe Next Door

A Journey Through 55 Alternative Realities, Parallel Worlds and Possible Futures

Edited by Frank Swain

First published in Great Britain in 2017 by John Murray (Publishers)
First published in the US in 2017 by Nicholas Brealey Publishing
An Hachette UK company

1

A CIP catalogue record for this title is available from the British Library and the Library of Congress

UK ISBN 978-1-473-62861-8
UK Ebook ISBN 978-1-473-62862-5
US ISBN 978-1-473-65867-7
US Ebook ISBN 978-1-473-65868-4

Typeset in CelesteST by
Palimpsest Book Production Limited, Falkirk, Stirlingshire

Printed and bound by CPI Group (UK) Ltd, Croydon CRO 4YY

John Murray policy is to use papers that are natural, renewable and recyclable products and made from wood grown in sustainable forests. The logging and manufacturing processes are expected to conform to the environmental regulations of the country of origin.

John Murray (Publishers)
Carmelite House
50 Victoria Embankment
London EC4Y ODZ

Nicholas Brealey
Hachette Book Group
53 State Street
Boston MA 02109

www.johnmurray.co.uk

www.nicholasbrealey.com

Contents

Welcome to
THE UNIVERSE NEXT DOOR

Imagine you could open a door and step into a different universe. It would be just like the one you live in, but with one important difference. In the first, the dinosaurs were never wiped out by a giant meteorite smashing into the Earth, and they still rule the Earth. In another, it's humans who walk the Earth – but there's no coal, oil or gas, so their eerily familiar society runs on wood, steam and elbow grease. And the universe over that way is the Earth of the far future, slowly unwinding under a dying sun. Does anything live there at all – and if so, is it animal or machine?

These imaginary universes, separated from our own by accidents of fate, gulfs of time or chasms of quantum weirdness, may seem the stuff of daydreams. And so they are. But thinking about them can amount to much more than amusing speculation, although you'll find a lot of amusement, and a fair bit of speculation, in the pages of this book.

A trip down these otherworldly rabbit holes begins with a basic proposition, then pursues all the questions that flow logically from it, no matter how ridiculous or counter-intuitive. Doing so, we arrive at startling realisations – not about the universes next door, but about our own.

That shouldn't really be too surprising. After all, asking questions is always the basis of how we understand the world around us. 'Why?' is perhaps the first question we ask, as inquisitive (and persistent) toddlers, trying to form a coherent model of the world around us. Many of us continue to ask it, for much the same reason, as we grow into adulthood. Some make a living from asking it, such as scientists and philosophers. Others are just fascinated by it, such as *New Scientist* readers.

'How?' is another excellent question. How do stars work? How does life work? And once you get a handle on the answers to those, you can start adding them together to make more complicated ones: how do stars make life work? The answers often turn out to be surprising and profound.

You can get a very long way by just asking how and why – from infant to grown-up, and from the basic evidence provided by our own senses to detailed models and powerful explanations for the universe in all its grandeur, intricacy and variety. But there are still plenty of big questions left to answer.

Are we alone in the universe? Do we really have free will? Where does our sense of self come from? When it comes to answering these, reductionist approaches – breaking down the problem into its elements and asking 'why?' or 'how?' over and over again – can be a slow and laborious way of forging understanding.

Enter a third kind of question, the kind celebrated in this book: what if? We know how things turned out in the universe we observe around us. But it can be hard to tell if they had to turn out that way, or if it's just chance that they did. So why not ask how things could have turned out differently?

So what if the dinosaurs hadn't died out? Would they still rule the Earth today?

How about if we all stopped eating meat? Would that really save the environment?

Or how about if you had simply decided to have something different for breakfast?

These kind of questions are valuable because they force

us to abandon basic assumptions about how the universe works. That helps us to separate accidents of fate from deep truths – and can lead to answers far more intriguing than our intuition would lead us to expect.

Dinosaurs might have continued to rule the Earth, but that doesn't mean they'd ever have developed intelligence like ours. Going vegetarian isn't the ecological panacea you might think. And your choice of breakfast, like every other decision you'll ever make, no matter how trivial, might spawn whole new universes.

Read on to find out our answers to all these and a great many other alternative realities, parallel worlds and possible futures – including some you've probably mused about, and quite a few you probably haven't. Welcome to the universe next door.

Sumit Paul-Choudhury
Editor-in-Chief, *New Scientist*

1 Playing Dice with the Universe

'There are more things in heaven and earth, Horatio,/Than are dreamt of in your philosophy.' So muses young Hamlet, after a troubling encounter with the ghost of his father. Hamlet's words are an apt description of the multiverse, the term physicists use to describe the infinite parallel universes that could exist alongside our own. Together, they provide more realities than we could dream of in a lifetime.

This shortfall is not for want of trying. In the four hundred years since Hamlet first took to the stage, we've continued to build imaginary universes in our heads, rearranging cogs of the great celestial clockwork just to see how it will affect the mechanism. Often, these strange and improbable worlds are created to challenge our own assumptions about how the universe works. Sometimes they're used to tease out truths about the universe itself.

Black holes, subatomic particles, time dilation and gravitational waves were all dreamt of in the philosophies of scientists long before we discovered evidence to prove their existence. In the multiverse, all these possible universes exist at once, so no matter how weird your imaginary universe is, there's one out there like that. It just might turn out to be the one you're living in.

Can we rewrite the laws of physics without destroying the universe?

? The universe seems fine-tuned for life. Is this evidence of a celestial designer, or the ultimate tautology? Michael Brooks explores how we can change the laws of physics without rendering the cosmos unfit for habitation.

You're a tourist visiting a grand country house. Wandering around, you notice that one of the rooms contains your ideal kind of reading chair. Then you see that the bookshelf next to it contains all your most-loved books, and a bottle of your favourite whisky sits on the side table. Is it a coincidence? Or did someone know you were coming and make the room just how you'd like it?

Scientists – cosmologists in particular – are asking a similar kind of question. Some have pointed out that certain natural features of the universe are peculiarly well suited to foster the emergence of life, and maybe even to facilitate the evolution of intelligent beings. Is that a coincidence? Is it evidence that the universe was set up with the eventual appearance of human life in mind? Or is it simply because our kind of life is likely to emerge in this kind of cosmos?

Such questions were first formalised by cosmologist Brandon Carter in 1973. Carter formulated two versions of what he termed the 'anthropic principle'. The weak version says that the properties of any universe we inhabit, along with its laws and contents, are probably restricted by the fact of our existence. In other words, for us to be here, the universe had to be just as it is. The strong anthropic principle is more controversial. It claims that the universe was constrained by the need for life to develop, a suggestion that invokes the idea of purposeful design.

Scientists have been interested in the anthropic principle ever since, partly because we have made some intriguing discoveries about the laws and constants of physics. Take the physical constant known as Omega, for example. This is the ratio of the energy density of the universe to the 'critical' energy density. The critical energy density allows a universe to expand from birth slowly enough for gravity to pull stars and planets together, but fast enough that it doesn't all pull back into a big crunch, destroying any possibility of life's emergence. How close is the actual energy density to this critical density? Very close. At the beginning of the universe, it was within 1 part in 10^{15}.

What else seems 'fine-tuned'? The efficiency with which hydrogen fuses to helium, which has to do with the strength of the attraction between particles in an atomic nucleus. It's about 0.007. Raise it to 0.008 and the hydrogen created in the big bang would have turned to helium almost immediately. Reduce it to 0.006, on the other hand, and helium would never form, the stars wouldn't have ignited to give energy for life.

Then there's the large discrepancy between the electromagnetic force and the gravitational force. This shapes the character of atoms in a way that a tiny change stops planets forming around stars or, going the other way, stops supernovae from creating the carbon atoms required for life as we know it. A 1 per cent reduction in the neutron mass prevents atoms from forming.

So numerous are these coincidences that the astronomer Fred Hoyle once suggested that the whole universe looked like a 'put-up job'. It's certainly possible to see them as evidence of a 'designer' who wanted to make a universe that would encourage the emergence of life. But could things be different and still give rise to life?

We are gathering hints that they could. Fred Adams of the University of Michigan at Ann Arbor, for example, showed in 2016 that the coincidence Hoyle identified – to do with the creation of carbon in supernovae – could happen by other means. A small tweak to the strong force that holds nuclei together would create conditions that make it easy to form carbon and other heavy elements. It's likely that such a tweak would produce other effects, but the proof of principle is there: you can change the laws of physics and still have the conditions for life emerge.

Adams has also shown that about a quarter of the possible values of the gravitational constant, nuclear reaction rates and the cosmological constant, which determines the expansion rate of the universe, still allow stars to form and burn. And in 2006, US physicists Roni Harnik, Graham Kribs and Gilad Perez showed that you could even have life in a universe entirely missing the weak nuclear force, one of the four fundamental forces.

In a 'weakless universe', stars still burn for billions of years, and chemistry and nuclear physics are 'essentially unchanged'. Interestingly, they also found that changing the cosmological constant does make a difference to the universe's ability to create large objects such as stars. That said, cosmologist Don Page at the University of Alberta in Edmonton, Canada, has argued that reducing the cosmological constant might actually create better conditions for the emergence of life.

There's another twist, too: the laws of physics may have changed over time. There are suggestions, derived from astronomical observations, that the fine structure constant, which determines the way light and matter interact, might have been different in the past. The ratio of the speed of gravity and the speed of light might also have changed since the big bang, according to some theorists. That would account for the 'horizon problem', a puzzle concerning how heat radiated through the early universe. There are even suggestions that the laws of physics could be different in different parts of the cosmos.

The biggest undoing of strong anthropic thinking came with the idea that ours was not the only universe. Many physicists are inclined to believe that the conditions that gave rise to our universe also cause new universes to 'bud off' from existing ones. Each of these universes can have slightly different laws of physics.

This idea of a multiverse solved a major problem faced by string theorists. Their equations give them multiple solutions: myriad universes with a diverse range of physical qualities. It was seen as a failing, until Stanford University's Leonard Susskind and others began to champion the multitude of solutions as a positive. Why shouldn't there be multiple universes, each with its own laws, Susskind asked. What's wrong with a bit of variety?

Recently, numerous physicists have begun to cluster around the notion that we are living in one of many possible universes, each of which could have its own discrete physical properties. Of course, we could only find ourselves living in one that had the right conditions for life to emerge. In other words, they say, anthropic thinking stems from an observer's selection effect.

That's all very well, but is there any evidence that such a multiverse actually exists? In 2007, cosmologists at York University in Toronto, Canada, suggested that colliding universes might leave an imprint on each other. This bruise would take the form of a hot, bright ring of photons. In 2015, Ranga-Ram Chary, who works at the US data centre of the European Space Agency's Planck telescope, spotted a set of unexplained bright patches on maps of the sky.

It's not conclusive evidence, and neither is that found by Tom Shanks at Durham University, UK. In April 2017, Shanks and his team published a paper looking at an anomaly in the cosmic microwave background. This sea of radiation was formed just after the big bang, and contains many clues to the state of the early universe.

Shanks's interest was piqued by a 'cold spot' in the cosmic microwave background. This feature is about 0.00015°C colder than surrounding areas. It may not sound like much, but it is enough to have people reaching for explanations. The most obvious one was that this area was a giant 'void' that contained relatively few galaxies. However, investigations by Shanks and his colleagues have ruled out this possibility. That, they say, leaves the door open for the cold spot to be the result of a collision with another cosmic bubble: the universe next door.

So where does all this get us? If you believe in the strong anthropic principle – essentially a deity-created universe perfectly suited to the evolution of humans – you don't have to change your views. The exact state of our universe is certainly compatible with a purposeful design hypothesis, although you would have to admit that there are other ways in which a designer could have done things. Just as there's no one way to build a car, there's no one way to build a life-friendly universe. If you're more of a weak anthropic person, you're also OK: yes, of course the universe will suit the observers within. And a different universe would suit different beings.

But if you're looking for concrete answers about why things are as they are, you'll be disappointed. Maybe the answer lies in another branch of the multiverse?

Mapping the multiverse: how far away is your parallel self?

? There seem to be an infinity of invisible worlds lurking out there. Finally we're starting to get a handle on where they are, and what it might take to reach them, says Shannon Hall.

Some of your doppelgängers mimic your every thought and action, only with a snazzier haircut. Some live in a world where the Nazis won the second world war, or where the dinosaurs survived, or where things fall up instead of down. Not here. Not in this universe. But they are out there – in the multiverse, where every possible world exists, along with all the infinite versions of you.

Travel any distance in modern fundamental physics and you will soon find yourself in the multiverse. Some of our most successful theories, from quantum mechanics to cosmic inflation, lead to the conclusion that our universe is just one of many. 'It's proven remarkably difficult to come up with a theory of physics that predicts everything we can see and nothing more,' says Max Tegmark.

So where are these unseen universes in relation to ours? How many are there? What goes on inside them? And can we ever hope to visit one? Such questions might sound daft, particularly given the lack of observational evidence that the multiverse exists. And yet thanks to new ideas on where distant universes might be hiding or how to count them, physicists are beginning to get their bearings. Rather fittingly, though, there is not just one answer – depending on which version of the multiverse you're navigating, there are many.

The journey into this confusion of worlds starts in our own. The universe we call home was born from the big bang some 13.8 billion years ago, during which time light has travelled farther than you might expect – 47 billion light years, thanks to the universe's ongoing expansion. This is the limit of what we can see, because light from more distant reaches would not have had time to reach us yet. But we're pretty sure space-time stretches farther, perhaps to infinity.

Past the cosmic horizon is a patchwork quilt of separate universes like ours, all bound by the same laws of physics. At least that's the assumption: those laws don't change over the distances we can see, so there is no reason to think they will suddenly transform beyond them. The only real differences are in the details: any intelligent life out there might live in a solar system that contains five planets instead of eight, say, or two suns.

What are the chances those details are exactly the same, to the point where there's another version of you? Tegmark thinks it's entirely plausible. Assuming space stretches on for ever, then there are an infinite number of patchwork universes, and everything allowed by the laws of physics will happen – more than once. He has even calculated the distance you would have to travel to meet your doppelgänger. In metres, it's a 1 followed by a hundred thousand trillion trillion zeros.

For many, though, this is just the first step. The expansion of the multiverse beyond the patchwork version started with a theory devised by Alan Guth in 1980. He proposed that in the first split second following the big bang, the early universe underwent a stupendous growth spurt, expanding by a factor of 10^{25}. This exponential ballooning, known as inflation, is beloved by cosmologists because it fixes several major problems with the big bang story.

In one guise, known as eternal inflation, the theory has space-time expanding exponentially for ever, but with quantum effects that stop the ballooning in small regions.

Our universe grew up in one of the resulting bubbles, and the same happened elsewhere. What's more, these quantum effects continue today, fuelling an endless froth of bubbles, each containing a universe.

Exactly where each bubble emerges is random. Picture a table of haphazardly placed snooker balls: one of those balls is our universe, and the others are additional universes separated by space-time. Then forget that picture. The analogy falls apart when you remember that each ball is growing as each universe expands, the table is stretching as the fabric between universes continuously inflates, and more balls are popping up at random as quantum effects spawn more universes. That's a spooky game of snooker.

Wherever they pop up, it's possible that these bubble universes – unlike those of the patchwork multiverse – contain physics gone wild. In 2000, Joseph Polchinski of the University of California at Santa Barbara and his colleagues threw string theory into the mix. The result gives rise to universes drastically different from our own, where unfamiliar laws of physics act on unimaginable particles. 'It's the multiverse on steroids,' says Alexander Vilenkin.

The reason is that string theory, a comprehensive but untested theory of nature, operates in ten dimensions – six more than the ones we know so well. These extra dimensions are scrunched up into unimaginably small spaces. In our universe, they form a particular configuration, which determines the properties of our particles and laws of physics. But they can form at least 10^{500} different configurations, which means an infinite number of universes can fall into that many categories, each with different particles and laws of physics. Basically, anything goes.

These bubble universes would be largely unrecognisable. Photons might outpace our speed of light, for instance, and apples might fall upwards from a tree's branch. They would also be inhospitable to us, because the stability of atoms depends on a certain balance of the constants in our theories, says Sabine Hossenfelder of the Frankfurt Institute for Advanced Studies in Germany. 'So if you go to another universe, it might also have areas of stability for atom-like things, but they would be very different from ours . . . Probably what would happen is that you would decay immediately.'

And even if there is another universe conducive to our sort of life, it would be well beyond reach – more distant than in the patchwork multiverse. 'You could never get there even if you travelled at the speed of light for ever because you would have to travel through a piece of space that is still inflating and doubling in size like crazy,' Tegmark says.

Or maybe there is a short cut. At least, there was. Last year, Vilenkin and his colleagues suggested that other universes might be nestled within the black holes that formed during the first second of our universe's existence. The idea is that small patches of space-time shifted into a different quantum state, forming tiny bubbles. Then, when inflation ended, those bubbles collapsed to form primordial black holes. But in the largest of these black holes, inflation continued, creating baby universes.

Vilenkin's theory predicts a distinctive distribution of black holes. Should this match the distribution of black holes in our universe, which we have only begun to chart, we'll have proof that this multiverse exists. We would even discover the mass a primordial black hole must have to contain another universe, potentially pointing us towards other universes dotted around the night sky.

We could never visit them, or could we? Nikodem Poplawski of the University of New Haven, Connecticut, and his colleagues also think that black holes harbour hidden universes, but these ones might still be attached to our own.

They came up with the idea in an attempt to revise Einstein's theory of general relativity, which indicates that there are 'singularities' at the heart of black holes that take up no space but are infinitely dense and infinitely hot. This aspect of the theory has always been hard to swallow. Even Einstein himself thought that singularities could not exist in physical reality.

If Poplawski is correct, they don't. According to his theory, the matter within a black hole doesn't collapse to a single point, but instead hits a barrier before bouncing back. 'But it cannot go back outside the black hole, which means the matter has to create a new space,' Poplawski says. 'So the black hole becomes a doorway to a new universe.'

The natural conclusion is that our universe was also formed within a black hole in another universe – an idea that sheds a rather different light on our beginning. 'The big bang is replaced by a big bounce,' says Poplawski. This origin explains our universe's expansion and does away with the need for inflation. In 2016, Poplawski even calculated that our parent black hole is probably a billion times the mass of the sun, on a par with the supermassive black holes that lurk in the centres of most massive galaxies.

That's not to say we should picture our universe as being within a universe, like a cosmic Russian doll. 'It's not the same physical space in some sense – it's more like a parallel universe,' says Damien Easson of Arizona State University in Tempe, who has made a similar speculation. 'It really would be a different universe altogether, occupying a different part of the multiverse.'

You might, however, be able to reach it via a short cut through space and time known as a wormhole. If this multiverse were represented by billiard balls, they would be connected by invisible tunnels, and they'd all be different sizes, growing at different rates. Some would form further black holes, which in turn would create more universes, themselves connected by invisible tunnels to those that created them.

With both the patchwork multiverse and the various bubbly versions, you can imagine them as contiguous – next to each other, or at least connected, even if the laws of physics in our universe mean you can't get from one to another. Not so for another sort, the worlds of the quantum multiverse. These are superimposed into the same space we occupy and are at once more intimate and more distant.

Since the 1920s, physicists have been baffled by quantum mechanics, which suggests that a particle can exist simultaneously in two or more possible states of being. So an electron, for example, can be in two places at once – until someone measures it. At that point, the electron has to 'choose' one particular state. But what about the other state?

In the 1950s, Hugh Everett, then a graduate student at Princeton University, came up with the idea that all of these potential states are equally real – they simply exist in parallel universes. Suppose, for instance, that you conduct an experiment in which you measure the path of an electron. In our universe, the electron travels in one direction, but the measurement creates another universe where the electron travels in the opposite direction.

In the quantum multiverse, then, every measurement creates another universe that is folded into our own yet is invisible and inaccessible.

That's too outlandish for some. The main problem, says Michael Hall of Griffith University in Brisbane, Australia, comes from confusion over what constitutes a measurement. Does it have to be a physicist doing a quantum experiment? Or could it be every decision we ever make? Everett's theory doesn't have the answers, so his many worlds remain fuzzy and the number of universes is impossible to count.

Fed up with the uncertainty, Hall and his colleagues have come up with a new scenario called 'many interacting worlds'. Unlike Everett's original idea, it begins with a finite number of universes, all similar to ours in size and scope, superimposed on our universe. Here, quantum events are produced as a result of particles from one of these universes interacting with those from another.

Take quantum tunnelling, where particles defy ordinary physics by tumbling through an energy barrier as though it isn't there. If an electron is heading towards a barrier in our world, it might interact with an electron heading towards that same barrier in a parallel world. The particles will start to repel one another, causing one to give the other an energy boost so that it can achieve the unimaginable: it will break through the energy barrier.

In Hall's theory, the probability of the electron breaking through is slightly different to the probability predicted by standard quantum mechanics. That's good news – in theory, the deviation is measurable and could tell us how many parallel universes we're dealing with.

But even if quantum many worlds do work in this way, how do they fit with the various cosmological multiverses? Hall thinks that his quantum multiverse and the inflationary multiverse can exist simultaneously, but argues that one has the upper hand. 'There is, in essence, only one super quantum multiverse,' he says. That's because inflating space-time creates new universes from small quantum effects, and in order for those effects to exist there must already be a quantum multiverse.

Others think that if more than one theory pans out, the inflationary multiverse probably came first. Once the frothing sea of inflation started churning up new bubble-like universes, a few would have created black holes, which formed more universes. Or perhaps those universes might branch off into other universes with every quantum measurement. 'If this interpretation is correct, then the eternally inflating multiverse will constantly split into a multitude of eternally inflating copies,' Vilenkin says.

Ultimately, there is nothing to say that all these different multiverses couldn't coexist. Some theorists even think that Everett's many worlds and the inflationary multiverse might actually be one and the same. Although they look different, they both constantly branch into new universes, says Leonard Susskind. And such an odd characteristic should not be taken lightly.

'Maybe they are different sides of the same story,' says Susskind. If so, the pioneers surveying the multiverse might finally agree that we only need one map to find our way.

Should you care about your parallel lives?

? Before setting off on your adventure through the multiverse, you might stop to consider the ramifications your actions have. Every decision you make may spawn parallel universes where people are suffering because of your choices. Rowan Hooper navigates the quantum moral maze.

I'm rich. I'm a movie star. I'm king of the world. I'm also poor. I'm homeless. Lots of me are dead. I'm none of these. Not in this universe. But in the multiverse I'm all of them, and more. I'm not a megalomaniac or a fantasist, but I do have a fascination with what-ifs. In the many-worlds interpretation of quantum mechanics, every decision I take in this world creates new universes: one for each and every choice I could possibly make. There's a boundless collection of parallel worlds, full of innumerable near-copies of me (and you). The multiverse: an endless succession of what-ifs.

In one of those worlds, I've just written a paragraph which explains that more clearly. This worries me. If many worlds is correct – and many physicists think it is – my actions shape the course not just of my life, but of the lives of my duplicates in other worlds. 'In the many-worlds interpretation, when you make a choice the other choices also happen,' says David Deutsch. 'If there is a small chance of an adverse consequence, say someone being killed, it seems on the face of it that we have to take into account the fact that in reality someone will be killed, if only in another universe.'

Should I feel bad about the parallel Rowans that end up suffering as a result of my actions? If I drive carelessly here, I might get a fine, but one of my other selves might crash and kill himself. Or worse, kill my parallel family. How am I supposed to live with the knowledge that I am just one of umpteen Rowans in the multiverse, and that my decisions reach farther than I can ever know? You might think I should just ignore it. After all, the many-worlds interpretation says I'll never meet those other versions of me. So why worry about them?

Well, most of us try to live by a moral code because we believe the things we do affect other people, even ones we'll never meet. We worry about how our shopping habits affect workers in distant countries; about as-yet-unborn generations suffering for our carbon emissions. Deutsch points out that we readily accept that attempted murder has moral implications, albeit less serious than actual murder. So why shouldn't we afford some consideration to our other selves?

Max Tegmark understands my quandary. A leading advocate of the multiverse, he's thought long and hard about what it means to live in one. 'I feel a strong kinship with parallel

Maxes, even though I never get to meet them. They share my values, my feelings, my memories – they're closer to me than brothers,' he says.

Taking the cosmic perspective makes it difficult for Tegmark to feel sorry for himself: there's always another Max who has it worse than him. If he has a near-miss while driving, he says he takes the experience more seriously than he did before he knew about the multiverse. 'The minimum tribute I can pay to that dead Max is to really think through what happened and learn some lessons.'

Once Tegmark would have been an outsider. When many-worlds was first proposed by Hugh Everett, it met with a scornful reception. Everett struggled to get it published, and eventually left academia in disgust. But its elegant explanations for some puzzling quantum phenomena have convinced more and more physicists over the past fifty years. 'Multi-universe physics has the same kind of experimental basis as the theory that there were once dinosaurs,' says Deutsch.

Nor can we avoid its consequences. Every time we make a decision that involves probability – such as whether to take an umbrella in case of rain – our decision causes the universe to branch, explains Andreas Albrecht at the University of California, Davis. In one universe, we take the umbrella and stay dry; in another, we don't, and we get wet. The fundamental variability of the universe forces such choices upon us. 'There's no escaping them,' says Albrecht.

That's a momentous realisation. We're living in a time akin to Copernicus realising that Earth wasn't at the centre of the universe, or when Darwin realised that humans were not created separately from the other animals. Both of those realisations reshaped our conception of our place in the universe, our philosophy and morality. The multiverse looks like the next great humbler of humanity.

'That these worlds are actually out there somewhere, but we cannot access them: I think that's an amazing and remarkable thing,' says physicist Seth Lloyd, a colleague of Tegmark's at MIT. 'It's sort of distressing really.' Why, I ask: because it diminishes humanity's status even more? 'No, not for that reason. I've always enjoyed the gradual marginalisation of humanity,' he says. 'No, it was somehow tidier to have the universe *be* the cosmos. But actually I'm liking it more, now that you've pointed out that it's really like the ultimate step in the marginalisation of human beings. I think that's much better. I enjoy that.'

Enjoyable though the multiverse might be as an concept, it's tough for us humans to get our heads around its implications – even for physicists themselves. When Tegmark's wife was in labour with Philip, their eldest son, he found himself hoping that everything would go well. Then he admonished himself.

'It was going to go well, and it was going to end in tragedy, in different parallel universes. So what did it mean for me to hope that it was going to go well?' He couldn't even hope that the fraction of parallel universes where the birth went well was a large one, because that fraction could in principle be calculated. 'So it doesn't make any sense to say "I'm hoping something about this number." It is what it is.'

Hope, it turns out, is the next casualty of the multiverse. You make a decision, and you end up on a branch of the multiverse with a 'good' outcome, or you find yourself on a 'bad' branch. You can't wish your way on to a good one. Tegmark acknowledges this is not easy to live with. 'It's tough to get your emotions to sync with what you believe,' he says. Too tough for me. How am I meant to live without hope?

What do other non-specialists make of the multiverse? When Hugh Everett died in 1982, aged just fifty-one, his teenage son Mark found his body. I asked him if his father's work had influenced him. 'Although I consider myself an Everettian by default, it's all beyond me for the most part, having inherited none of my father's mathematical smarts,' says Mark, long-time frontman of the band Eels. 'How can I grasp any of it except in small moments? I'm having a hard enough time dealing with this world lately. I only hope some of the other worlds are easier for me to figure out.'

I know how he feels. Perhaps a philosopher can help me take a broader perspective. I turn to David Papineau of King's College London. 'Say you put your money on a horse which you think is a very good bet,' he says. 'It turns out that it doesn't win, and you lose all your money. You think, "I wish I hadn't done that." But you brought benefits to your cousins in other universes where the horse won. You've just drawn the short straw in finding yourself in the universe where it lost. You didn't do anything wrong. There's no sense that the action you took earlier was a mistake.'

Hmm. I doubt 'I didn't make a mistake' would get much traction with my partner if I bet all our savings on a horse this afternoon and find myself on the 'wrong' branch. But then, that wouldn't be the sensible thing to do – and one of the great attractions of Everett's interpretation, according to Papineau, is that it's not 'messy', as long as you act rationally.

With orthodox thinking, there are two ways of evaluating risky actions, he says. First, did you make the choice that was most in line with the odds? If we needed money, and my stake had been proportionate, it might have been. Second, did it work out well? There are any number of reasons it might not – the horse might fall, or just defy the odds and trail in last.

It offends Papineau that these two ways of being 'right' – choosing wisely and getting lucky – don't go hand in hand. 'The idea that the right thing to do might turn out to have been the wrong thing seems to me to be a very ugly feature of orthodox thinking,' he says. This doesn't arise in the many-worlds interpretation, where every choice is made and every outcome occurs. That leaves no place for hope or luck, but nor does it leave room for remorse. It's an elegant, if cold-blooded, way to look at things.

This elegance has always been part of the multiverse's appeal. In quantum mechanics, every object in the universe is described by a mathematical entity called a wave function, which describes how the properties of subatomic particles can take several values simultaneously. The trouble is, this fuzziness vanishes as soon as we measure any of those properties. The original explanation for this – the so-called Copenhagen interpretation – says the wave function collapses to a single value whenever a measurement is made.

Hugh Everett called this enforced separation of the quantum world from the everyday, classical one a 'monstrosity', and decided to find out what happened if the wave function did not collapse. The resulting mathematics showed that the universe would split every time a measurement is made – or in human terms, whenever we make a decision with multiple possible outcomes. That's the many-worlds interpretation.

For theoretical physicist Don Page, this elegance goes far beyond human actions. Page is both a hard-core Everettian and a committed Christian. Like many modern physicists, he agrees with Everett's stance that collapsing the wave function is unnecessarily complicated. What's more, for Page it has a happy side effect: it explains why his God tolerates the existence of evil.

'God has values,' he says. 'He wants us to enjoy life, but he also wants to create an elegant universe.' To God the importance of elegance comes before that of suffering, which, Page infers, is why bad things happen. 'God won't collapse the wave function to cure people of cancer, or prevent earthquakes or whatever, because that would make the universe much more inelegant.'

For Page, that is an intellectually satisfying solution to the problem of evil. And what's more, many worlds may even take care of free will. Page doesn't actually believe we have free will, because he feels we live in a reality in which God determines everything, so it is impossible for humans to act independently. But in the many-worlds interpretation every possible action is actually taken. 'It doesn't mean that it's fixed that I do one particular course of action. In the multiverse, I'm doing all of them,' says Page.

There are limits to Page's willingness to leave his fate to the multiverse, however. Seth Lloyd once offered him $1 million to play quantum Russian roulette (see below), which is a good game for a multiverse aficionado: you can't lose. Page thought about it, then declined: he didn't like the thought of his wife's distress in the worlds where he died.

Like Tegmark, Page seems to value the multiverse for the perspective it offers. 'One of my teenage children wants to get a motorcycle, which my wife and I think is pretty dangerous,' Page says. 'But if I say: "OK, maybe most of the time you'd survive, but there's going to be some part of you, some branch, in which you get seriously maimed in a motorcycle accident" . . . Maybe I'll try it.'

I'm somewhat relieved to find that even many-worlds experts ultimately behave in much the same way as people who know nothing of it. But I've also realised that it shapes the way they think about their decisions. Perhaps it's more natural for us to think about how our actions affect our 'other selves' than about the arid probabilities of risk and reward. If anyone's going to buck this trend, it's surely David

Deutsch, probably the most hard-core of Everettians. Surely he can give me the last word on what it means to live in the multiverse. He does, but it is by no means the answer I was expecting.

'Decision theory in the multiverse tells us that we should value things that happen in more universes more, and things that happen in fewer universes less,' he says. 'And it tells us that the amount by which we should value them more or less is, barring exotic circumstances, exactly such that we should behave as if we were valuing the risks according to probabilities in a classical universe.' So the right thing to do remains the right thing to do.

So has my quest been for nothing? Not at all. For one thing, Deutsch's approach could be wrong, a possibility he accepts, though he is adamant the multiverse exists. But if he's right, his conclusion only reinforces what his peers have been telling me: the best way to live in the multiverse is to think carefully about how you live your life in this one.

Thinking of what-ifs as having some kind of reality can help us to do that. Tegmark says many worlds has made him think differently about life. He sometimes fears doing something because it feels too big a deal. But then he realises that in the grander context of the multiverse, it's not big at all – and he just does it. 'The multiverse has definitely made me a happier person,' he says. 'It's given me courage to take chances to be bold in life.'

I hope it will do the same for me. We might not stop feeling hope or remorse, but the multiverse can help put those feelings in perspective. And while the multiverse may not require a change in our morality, it can help us think harder about our choices and actions. The cosmos reaches far farther than we ever appreciated. But so, it seems, do we.

How to play quantum Russian roulette

This amounts to playing the role of Schrödinger's cat. You'll need a gun whose firing is controlled by a quantum property, such as an atom's spin, which has two possible states when measured. If the Copenhagen interpretation is right, you have the familiar fifty-fifty odds of survival. The more times you 'play', the less likely you are to survive.

If the multiverse is real, on the other hand, there always will be a universe in which 'you' are alive, no matter how long you play. What's more, you might always end up in it, thanks to the exalted status of the 'observer' in quantum mechanics. You would just hear a series of clicks as the gun failed to fire every time – and realise you're immortal. But be warned: even if you can get hold of a quantum gun, physicists have long argued about how this most decisive of experiments would actually work out.

Could there be a doorway to the multiverse in our backyard?

? Falling into a black hole may not be as final as it seems. Apply a quantum theory of gravity to these bizarre objects and the all-crushing singularity at their core disappears. In its place is something that looks a lot like an entry point to another universe. Most immediately, that could help resolve the nagging information loss paradox that dogs black holes, discovers Katia Moskvitch.

Though no human is likely to fall into a black hole any time soon, imagining what would happen if they did is a great way to probe some of the biggest mysteries in the universe. Most recently this has led to something known as the black hole firewall paradox – but black holes have long been a source of cosmic puzzles.

According to Albert Einstein's theory of general relativity, if a black hole swallows you, your chances of survival are nil. You'll first be torn apart by the black hole's tidal forces, a process whimsically named spaghettification.

Eventually, you'll reach the singularity, where the gravitational field is infinitely strong. At that point, you'll be crushed to an infinite density. Unfortunately, general relativity provides no basis for working out what happens next. 'When you reach the singularity in general relativity, physics just stops, the equations break down,' says Abhay Ashtekar of Pennsylvania State University.

The same problem crops up when trying to explain the big bang, which is thought to have started with a singularity. So in 2006, Ashtekar and colleagues applied loop quantum gravity to the birth of the universe. LQG combines general relativity with quantum mechanics and defines space-time as a web of indivisible chunks of about 10^{-35} metres in size. The team found that as they rewound time in an LQG universe, they reached the big bang, but no singularity – instead they crossed a 'quantum bridge' into another, older universe. This is the basis for the 'big bounce' theory of our universe's origins.

In 2013, Jorge Pullin at Louisiana State University and Rodolfo Gambini at the University of the Republic in Montevideo, Uruguay, applied LQG on a much smaller scale – to an individual black hole – in the hope of removing that singularity too. To simplify things, the pair applied the equations of LQG to a model of a spherically symmetrical, non-rotating 'Schwarzschild' black hole.

In this model, the gravitational field still increases as you near the black hole's core. But unlike with previous models, this doesn't end in a singularity. Instead, gravity eventually reduces, as if you've come out the other end of the black hole and landed either in another region of our universe, or another universe altogether. Despite only holding for a simple model of a black hole, the researchers – and Ashtekar – believe the theory may banish singularities from real black holes too.

That would mean that black holes can serve as portals to other universes. While other theories, not to mention some works of science fiction, have suggested this, the trouble was that nothing could pass through the portal because of the singularity. The removal of the singularity is unlikely to be of immediate practical use, but it could help with at least one of the paradoxes surrounding black holes, the information loss problem.

A black hole soaks up information along with the matter it swallows, but black holes are also supposed to evaporate over time. That would cause the information to disappear for ever, defying quantum theory. But if a black hole has no singularity, then the information needn't be lost – it may just tunnel its way through to another universe.

What if time flows backwards in some universes?

? Why does time seem to have a preferred direction? In the multiverse, pocket universes could be born with clashing directions of time – the evolving future of one could happen in the rewinding past of another. Joshua Sokol investigates.

Don't pity those in the past – in their own way, they might have a lot to look forward to. From our perspective, events in some universes may seem to unfold backwards. That implies there could be alternate worlds whose future is actually even farther in our distant past.

This trippy idea has been suggested before, often with very specific caveats. In 2004, Sean Carroll, now at the California Institute of Technology in Pasadena, showed it could apply, but only if complex and unlikely physics was involved. Carroll and cosmologist Alan Guth have since shown how time itself can arise organically from simpler principles, then flow in opposite directions in adjacent universes.

Guth and Carroll's work is motivated by a problem vexing physicists and philosophers: why it is that time's arrow points in just one direction. It's true we can only remember the past, but the laws of physics don't much care which way time flows: any physical process run backwards still makes sense according to those laws. 'There's no such thing, at a very deep level, that causes [must] precede effects,' says Carroll.

In the absence of other laws to set the direction of time, physicists have settled on entropy – basically, a measure of messiness. As entropy grows, time ticks forward. For example, you can stir milk into coffee but you can't stir it back out again – so neatly separated black coffee and milk always comes first.

Zooming out to the entire universe, we likewise define the future as that direction of time in which entropy increases. By studying the motion of faraway galaxies, we can predict how the cosmos will evolve. Or we can rewind time back to the big bang, when the universe must have had much less entropy.

Try to rewind farther and we meet a cosmological conundrum. We can't proceed if the big bang was indeed the beginning of time, but in that case, why did it have such low entropy? And if it wasn't the beginning of time – as Guth suspects – we'd still want to know how an eternal universe could have reached such a low-entropy state that would allow for the arrow of time to form.

In an as yet unpublished model, Guth and Carroll explore the latter idea. They drop a finite cloud of particles, each zipping around with its own randomly assigned velocity, into an infinite universe. After a while, arrows of time emerge spontaneously.

The random starting conditions mean that half the particles initially spread outwards, increasing entropy, while the other half converge on the centre, decreasing entropy, then pass through and head outwards. Eventually the whole cloud is expanding, and entropy is rising in tandem. Crucially, this rise happens even if you reverse time by flipping the starting velocity of every particle: ultimately, all particles will end up travelling outwards. If entropy grows either way, who's to say which way the arrow of time should point? 'We call it the two-headed arrow of time,' Guth says. 'Because the laws of physics are invariant, we see exactly the same thing in the other direction.'

The model shows that an arrow of time arises spontaneously in an infinite, eternal space. Since this allows entropy to grow without limit, time zero could simply be the moment where entropy happened to be at its lowest.

That could explain why the big bang, the earliest moment we can see, has so little entropy. But it also feels a little like a cheat: if entropy can be infinite, anything can have relatively 'low' entropy by comparison.

'The point that Alan and I are trying to make is that it's very natural in those circumstances that almost everywhere in the universe you get a noticeable arrow of time,' Carroll says, though he admits the model still needs work. 'Then of course you do the work of making it realistic, making it look like our universe. That seems to be the hard part.'

If the model matches reality, it would have implications for more than just our own observable universe. 'This is intended to describe the whole of existence, which would mean the multiverse,' Guth says. In his view, the arrow of time may have arisen in a parent or grandparent universe of our own.

According to cosmic inflation – an account, pioneered by Guth, of the universe's rapid early expansion – our cosmos is just one among many. Fresh pocket universes, next to ours but far enough away to be unreachable, spring from the vacuum all the time.

That would mean that the big bang represents the instant our universe sprang into being from the multiverse, after the arrow of time had already been set.

Parallel universes, born around the same time as ours, would have started with similar entropy to ours. If we could talk with beings there, they would agree with us on the direction of past and future.

But from our viewpoint, time would be turned on its head in universes that arose before our arrow of time was set. Nobody would notice, though. While small, random differences between these backwards universes and ours might lead to vastly different fates, living beings there would see an arrow of time, but 'to them we would be in the past and we would have the wrong arrow of time direction', Carroll says. It would be hard to have that argument, though. 'We can't talk to them; they are in our past. And they can't talk to us, because we are in their past,' he says.

The new model has its issues. Its initial state, when all particles are given random velocities, is fuzzy in that the diverging arrows of time are not yet clearly defined, since entropy is growing in some places while shrinking in others.

Understanding this period in between the two emergent arrows of time is hard. 'That nebulous region in the middle might turn into a real monster,' says Andreas Albrecht at the University of California, Davis.

It's a weakness Guth acknowledges – and along with the difficulty of incorporating elements like gravity, it's a reason why their work hasn't been published yet, he says. 'What we'd like to understand better is how to describe this middle region,' Guth says. 'But all of physics is described in terms of a system where the arrow of time is well defined, so it's a stretch to figure out how physics would behave.'

Albrecht also isn't sold on how the model handles an infinite universe, in which entropy can just keep growing for ever. That key feature explains why the big bang had 'low' entropy compared with now, but it's controversial.

'They've created a world where they can slip certain notions in very easily,' Albrecht says. He believes their use of infinite universes helps to hide big assumptions in the model. But Albrecht likes the possibility of two arrows of time, where the multiverse's distant past is also its far future. 'The double-headedness is something I totally embrace,' he says.

Could a theory of everything reveal the multiverse?

? The best dreams of our philosophy only describe a fragment of our observable universe. Might we one day be able to dream big enough to encompass the sum total of reality? wonders Richard Webb.

Let's not kid ourselves: everything we think we know now is just an approximation to something we haven't yet found out. That is the frustrating, exhilarating lesson history teaches us about fundamental theories of nature. Take Newton's universal law of gravity. It did sterling service describing falling apples and orbiting planets for over two centuries, but eventually gave way to a 'righter' theory – Einstein's general relativity. Ditto the solidly intuitive outlines of classical mechanics: dig down to the level of subatomic particles, and we find them fogged in a haze of quantum uncertainty.

Quantum theory explains matter's small-scale workings. General relativity describes the universe's large-scale evolution. Each theory is very right in its own way, but has omissions and inconsistencies that convince us that they, too, are just placeholders for something better. A unified 'theory of everything' would take us to places where quantum theory and relativity break down – beyond a black hole's event horizon, for example, or the very first instants of the universe.

Heady stuff – and many a great mind has come a cropper in the chase. Einstein's twilight years were largely spent in an isolated, fruitless quest for ultimate enlightenment.

Things haven't moved on much. String theory, which seeks unification by replacing fundamental particles with tiny scrunched-up strings, has come – and not yet gone, despite much criticism of its lack of testable predictions. Rival approaches such as loop quantum gravity have sprung up, but also brought no breakthrough.

An ill wind for the next sixty years? Mike Duff of Imperial College London, who has been studying theories of everything since the late 1960s, is surprisingly chipper about how much longer it will take. 'I'm sceptical of making predictions,' he says, 'but if pressed, I'd say it will take more than ten years, but less than a hundred.'

Carlo Rovelli of Aix-Marseille University in France, a leading light in loop quantum gravity, is also optimistic. 'Sixty years in the future we will have a theory of quantum gravity, in my opinion,' he says. If so, that will be a beginning, not an end. Theories earn credibility only gradually. Some predictions of general relativity, such as black holes and gravitational waves, were fully appreciated and explored only in the 1960s. The existence of gravitational waves was finally confirmed in 2016.

Complicating matters for a unified theory is that it will kick in at insanely high energies far beyond our wit to reproduce. Experimental proof, if any is possible, is likely to be subtle and indirect, perhaps taking the form of patterns imprinted in the cosmic microwave background – radiation left over from the big bang – for example.

For similar reasons, immediate practical benefits are unlikely – although never say never, says Rovelli. 'I don't see technological applications of the theory right now,' he says. 'But nobody imagined GPS back in the 1970s when I started studying general relativity.'

So don't expect any eureka moment – more a long, slow evolution. But equally, expect the unexpected. When Paul Dirac unified quantum theory and Einstein's special relativity in the 1920s to provide a full description of the electron, he ignored his equation's prediction of a second, almost identical particle. The discovery just a few years later of the positron opened our eyes to a whole new world that we have yet to explore fully – that of antimatter.

A theory of everything based on string theory, say, might prove the existence of a 'multiverse' of further universes, vastly increasing the scope of what we can hope to know – or not. That's all part of the fun of the chase, says Duff. 'Most scientists are trying to become masters at the game,' he says. 'Theoretical physicists are still trying to understand the game.'

An argument over the colour of light almost dashed any chance of a theory of everything. Turn to page 70 to find out why.

2 You're Not Here

The multiverse offers us an infinite variety of realities, but that doesn't mean they have to be impossibly exotic. Most of us have wondered how things might have turned out differently: if, say, we'd never accepted that first job, or hadn't plucked up the courage to speak to our current partner. In this chapter, we trace eleven alternative timelines, exploring how history as we know it might have taken a different direction were it not for one key event. Would we have a theory of evolution if Darwin had done what his father wanted and stayed home to become a parson? Would our civilisation have developed so far without an abundance of fossil fuels to tap?

Chaos theory tells us that arresting the flap of a butterfly's wings could prevent a future hurricane. But is the course of human history more resilient to such tiny changes? These are not idle musings – the answers explore everything from how fragile our scientific philosophies are to where might be the best place in the galaxy to find intelligent life.

What if Earth didn't have a moon?

? Could life have evolved on Earth without a moon? Or two moons, or didn't spin, or was a moon itself? Hazel Muir explores the science of six alternate Earths.

In the quirky H. G. Wells tale 'The Man Who Could Work Miracles', a character called George Fotheringay discovers that he has supernatural powers. Egged on by the local vicar, Fotheringay uses his gift to miraculously improve his village by night, mending buildings and reforming drunks. Then he realises there is a way to buy more time for good deeds before sunrise: simply order Earth to stop spinning.

The moment Fotheringay gives the command, all hell breaks loose. 'When Mr Fotheringay had arrested the rotation of the solid globe, he had made no stipulation concerning the trifling movables upon its surface,' Wells wrote. 'Every human being, every living creature, every house and every tree – all the world as we know it – had been so jerked and smashed and utterly destroyed.'

Wells loved to play God with the planets in his fiction, but such fantasies are good for more than entertainment. Implausible scenarios of the sort Wells cooked up are worth exploring because they give clarity to what makes the real world tick, says Neil Comins, an astrophysicist at the University of Maine in Orono. In his book *What If the Earth Had Two Moons?* Comins gives our planet some otherworldly makeovers to see how different our environment could have been. 'Thinking through what could happen or might have happened gives a much better perspective on how things actually are,' he says.

So what strange scenarios would drastically alter our world, and what would be the upshot?

Toying with the moon is a good place to start. The moon has had an enormous influence on our planet, not least in the cataclysmic events that led to its formation. The consensus is that a Mars-sized body, often called Theia, slammed into Earth around 4.5 billion years ago, spraying debris into orbit. This eventually clumped together to form the moon at about a tenth of its current distance, roughly the same altitude as most of today's communications satellites. Around this time, Earth would have rotated on its axis once every eight hours or so, but gravitational interactions between Earth and moon, including effects to do with the tides, have since slowed that to the familiar twenty-four-hour day.

So what if the moon had not formed? The only significant tidal forces on Earth would come from the sun, which would have increased the length of a day from eight to about twelve hours. You would weigh less too, since about 10 per cent of the mass of Earth is thought to come from the remnants of Theia that it absorbed, so gravity would be that much weaker.

Without the moon, life might not have taken hold as quickly as it did. The newborn moon was so close to Earth that it would have raised tides a thousand times higher than today's. Those vast tides probably caused the oceans to scour the continents, enriching the sea with minerals and helping to create the 'primordial soup' that gave rise to life. Comins suspects life would eventually have emerged even without the moon, but says there would not have been animals adapted to live in tidal pools, or to hunt or navigate by moonlight. What's more, with no lunar gravity to stabilise Earth's rotation, our planet could have ended up spinning on its side, like Uranus. Over the course of a year, sunlight would drift from one pole to the other, then back again. 'Virtually every living thing would have to migrate on such a world,' says Comins. 'Life would have to follow the sunlight.'

The impact of Theia could, in theory, have created more than one moon. Would that have made a difference? Probably not. Even if the debris had formed two lumps, gravitational effects would have made them collide long before complex life forms appeared on Earth some 600 million years ago.

The only way Earth could support a second moon today

– let's call it Moon2 – would be if it had been captured recently from a pair of bodies that wandered into its vicinity. This encounter could have left Moon2 settled into a stable Earth orbit, as long as a large chunk of its kinetic energy was transferred to its companion, which would have zoomed off into space. The gravity of Moon2 and its companion would have caused havoc as they approached Earth, triggering mammoth tidal waves and volcanism. The skies would have been dark with dust, and there would almost certainly have been a mass extinction of life. Things would have calmed down eventually – probably within a few years of the departure of Moon2's companion.

Suppose that Moon2 was the same size as the original moon and that its orbit was in the same plane and direction, but half as far away from Earth. Any surviving land dwellers would be treated to a spectacular second moon twice as wide and four times as bright as the first, circling Earth once every ten days. When both moons were full, it would be easy to read a book at midnight.

Not that it's all good news: Moon2 would be spitting lava. The ever-changing gravitational forces attributable to Earth and our original moon would massage Moon2's interior, keeping it molten and making lava spew out through volcanoes and cracks in the surface. 'How spectacular that would be,' says Comins. 'You'd also see glowing rivers of lava on this moon.' Some would be ejected fast enough to escape and fall to Earth, making clear nights a glittering show of shooting stars.

The two moons would be destined to collide. While tidal interactions make our moon recede from Earth by 3.8 centimetres each year, Moon2 would recede faster, catching up with the original roughly 1.5 billion years after capture. Their catastrophic collision would send debris raining down on Earth, likely causing another mass extinction.

Another of Comins's scenarios has a moon orbiting Earth the 'wrong' way, rather than in the same direction as Earth rotates. Given the manner of our moon's birth, that would be impossible: if Theia had carried enough momentum to splatter a moon into orbit in the opposite direction, it would almost certainly have destroyed Earth in the process, Comins argues.

So a moon orbiting in the opposite direction – called Noom, say – could only exist if it had been captured from a pair of passing bodies. According to Comins, that is unlikely, but not impossible. Let's say Noom has the same mass as our moon and orbits at the same distance with the same period, albeit in the opposite direction. The previously moonless Earth would now have a faster rotation rate, turning once in twelve hours.

The complex gravitational interplay between the two worlds would lead to Noom gradually spiralling towards Earth, its orbit becoming ever faster and possibly more elliptical. Meanwhile, Earth's rotation rate would slow to zero, before it eventually started spinning in the opposite direction.

As Earth's rotation ground to a halt, the days would become as long as the planet's year, causing extreme heating on the daylight side and extreme cooling elsewhere. But the slowing would occur over billions of years, so animals would have time to evolve migratory patterns to follow comfortable climates. 'There could easily be life on the boundary, where the sun is on the horizon,' says Comins.

After that, the sun would rise in the west and set in the east as Earth began to spin in the opposite direction. As Noom got closer, shorelines would be ravaged by tides reaching 3 kilometres high. Eventually, Noom would get so near that it would be torn apart by gravitational tidal forces, disintegrating into a ring of boulders around 4,500 kilometres above Earth.

Some of these would give the planet a good pummelling – perhaps severe enough to cause a mass extinction.

So much for monkeying around with the moon. But what if Earth were not a planet but a moon akin to the Earth-like moon Pandora in the film *Avatar*? Imagine Earth in orbit above the equator of a clone of Neptune – Neptune2 – and that both bodies rotate about an axis perpendicular to the plane of the solar system. For this satellite Earth to be warm and habitable, the system would have to orbit the sun at roughly the same distance as the real Earth does now.

Over a few billion years, the satellite Earth's rotation would become synchronised with its orbit, so that one face would be permanently turned towards Neptune2. If the satellite Earth were orbiting around 300,000 kilometres from the centre of the planet, it would have an orbital period and a day lasting just over one hundred hours. The view of Neptune2 from Earth would be spectacular, spanning 9 degrees of sky, or eighteen times the angular size of the full moon.

If you lived in the centre of the Neptune2-facing side, the planet would be directly overhead, and half of it would be lit up when the sun rose. It would shrink to a crescent before eclipsing the sun for about two hours around noon, the stars appearing in an inky black sky. Then Neptune2 would gradually grow through another crescent phase as night fell to a 'full Neptune2' around midnight, shining about 2,800 times brighter than our moon ever gets. Midnight on this side of Earth would be far brighter than noon. 'Effectively, there would be two periods of daylight,' says Comins. Animals living on the Neptune2 side and the far side would experience different day–night cycles and would have different body clocks.

The long days and nights on the satellite Earth would create daily temperature swings roughly twice as big as those on our Earth so life would have to adapt. Worse, Neptune2's gravity would make it a magnet for asteroids and comets and the captive Earth would be at risk from cosmic crossfire. 'Neptune2 would pull debris onto it, and that is also going to potentially threaten the Earth,' Comins says.

Now imagine drastically altering conditions on Earth, not by tinkering with the moon or having Earth orbit another planet, but simply by giving Earth a thicker crust. The average thickness of Earth's continental crust is about 40 kilometres, while the oceanic crust is 7 kilometres. What would the world be like if the crust was, say, about 100 kilometres thick on average? That could have come about if the young Earth had been very dry.

Most of the water on Earth is thought to have been delivered by icy comets and asteroids. This water makes Earth's crust and upper mantle – its lithosphere – flexible enough to be pushed aside when blobs of magma carry heat up from the planet's interior. 'Water provides lubrication for the movement of the crust,' says Comins.

Had comets brought water to Earth much later, its crust would have become much thicker. That's because, over time, magma blobs would have stacked up and congealed beneath the lithosphere. Heat would build up in Earth's interior until, eventually, something would give, causing parts of the lithosphere to melt all the way to the surface every few tens of millions of years. As these tracts of land or seabed turned molten, they would release heat to space over a few hundred years before solidifying again.

These meltdowns would release a cocktail of toxic gases and erase all features in the vicinity. 'Perhaps thousands of square kilometres are going to become totally uninhabitable,' says Comins, who based his model on the parched crust of Venus, which lacks tectonic plates. 'Venus has seen these melts, which is why the planet has so few craters – it has been resurfaced.' He adds that any successful species on Earth would need the ability to sense when the land under its feet was about to melt, and flee: 'Otherwise it's just going to be plain wiped out.'

Finally, back to the miracle worker in H. G. Wells's tale: what would happen if Earth suddenly stopped spinning? Certainly everything on the surface would continue moving at up to 1,667 kilometres per hour, the rotation speed at the equator. 'Anything on the surface that is not held down with incredible strength is going to fly off parallel to the surface,' Comins says. He calculates that people outdoors would be flung outwards to an altitude of about 11 kilometres, then fall and hit the ground at more than 1,000 kilometres per hour. Buildings would be ripped from their foundations, while the oceans would engulf the land. Such a catastrophe could extinguish all life on Earth.

Life would fare better if Earth stopped spinning over a longer timescale, say two or three decades. There would be a profound effect on the oceans, however. The centrifugal effect of Earth's rotation makes the solid Earth bulge outwards at the equator, and creates equatorial ocean bulges 8 kilometres high. Were Earth to stop spinning, the oceans would migrate to the poles, where surface gravity is slightly stronger because the land is closer to Earth's centre.

Witold Fraczek of the Environmental Systems Research Institute in Redlands, California, has simulated this scenario and shown that once Earth had lost half its angular momentum, the oceans would split into two parts, one at each pole, with shorelines at roughly 30 degrees north and south. In between, a mega-continent would emerge, its mountains peaking at up to 10 kilometres above the new sea level. The northern ocean would drown most of Canada, Europe and Russia.

Whether people could survive on this new world is unclear. Much of the agricultural land would be lost, and the atmosphere would become too thin above most of the equator for people to survive there. Humans would separate into two populations, living along the shores of the northern and southern oceans, kept apart by rugged terrain in between, Fraczek says.

Add to that the challenge of scorching days and frigid nights, each lasting six months, spring accompanying sunrise, sunset signalling autumn. People might live in the twilight zones, migrating to keep pace with the gradual shift of light around the globe.

While there's no way Earth could, in reality, stop spinning over just a couple of decades, its rotation is gradually slowing. Many billion years into the future, it's possible that Earth's day could become as long as its year. Wells pictures a perpetual sunset on the ageing Earth in his classic book *The Time Machine*, but beyond fiction, that future is just too distant to foretell.

Want more celestial rearrangements? Jump to page 236 to find out how Earth might escape a dying sun.

What if the dinosaurs hadn't been wiped out?

? Without the help of an asteroid strike, mammals might never have had their break, says Colin Barras. But granted a stay of execution, could dinosaurs have evolved intelligence like ours?

It's the most famous extinction event of them all. Roughly 66 million years ago, an asteroid slammed into what is now the Gulf of Mexico, leading to the demise of all the dinosaurs apart from the birds. The catastrophe helped give mammals their big break, ultimately paving the way for the evolution of upright apes. As such, humanity may owe its existence at least in part to the Chicxulub asteroid impact. But that hasn't stopped some biologists from wondering what the world might look like today if the giant space rock had sailed harmlessly past our planet all those years ago.

Running this sort of thought experiment isn't easy. 'Evolution is a chaotic, nonlinear process,' says Doug Robertson at the University of Colorado, Boulder. 'Prediction is essentially impossible in chaotic systems.' But the predictions that evolutionary biologists make fall on a spectrum, and those at one end are fairly secure.

Take, for instance, the likelihood that the non-bird dinosaurs would still exist in the twenty-first century had they dodged the asteroid. The fossil record tells us that non-bird dinosaurs dominated ecosystems for roughly 160 million years. 'They were still thriving, still hugely diverse when the asteroid hit,' says Steve Brusatte at the University of Edinburgh, UK. With such a track record of success, it's fairly likely that the non-bird dinosaurs would have continued to flourish to this day if the asteroid had missed, says Robertson.

Brusatte agrees with this prediction. He says we shouldn't ignore the fact that plenty has happened during the last 66 million years – global temperatures have dropped considerably, landscapes have change dramatically with the appearance of grasslands, and there have been other environmental crises. But Brusatte thinks the non-bird dinosaurs probably had what it takes to cope with these challenges. 'If the asteroid never hit, then the non-bird dinosaurs would still be here today, and they would probably be very diverse,' he says.

Another reasonably solid prediction relates to the likelihood of humans evolving in a world that still contained a multitude of non-bird dinosaurs. Our furry ancestors could certainly cope with life in a dinosaur-dominated world – the fossil record reveals that mammals thrived and diversified long before the asteroid impact. But for as long as the non-bird dinosaurs were around, mammals remained fairly small. If those dinosaurs hadn't vanished, mammals would probably still be diminutive.

'I doubt that there would have been any of the large-sized mammals that started appearing several million years after the end-Cretaceous extinction event,' says Hans-Dieter Sues at the Smithsonian Institute in Washington, DC. Primates might still have appeared, says Sues – most have small bodies – but large-bodied apes are unlikely to have succeeded. Consequently, there would probably be no humans in this alternate reality.

But a dinosaur-dominated world would not necessarily have been free of superintelligent life. At the shakier end of the prediction spectrum is the suggestion that some non-bird dinosaurs might have developed intelligence on a par with ours.

The idea is not entirely groundless. For instance, fossils reveal interesting evolutionary trends in the anatomy of some non-bird dinosaurs. 'The latest "models" were considerably more advanced in terms of brain size, diversity, limb length and just about everything else,' says Philip Currie at the University of Alberta in Canada.

In the 1980s, a palaeontologist called Dale Russell focused on one small predatory dinosaur that lived across what is now North America just before the asteroid struck. *Troodon*, noted Russell, had an unusually large brain relative to its body size. It was certainly no intellectual heavyweight – probably more on a par with an average bird – but it still stood out among the non-bird dinosaurs.

Russell calculated that, if the *Troodon* lineage had been

given the opportunity to continue down the path towards larger brains, by now it could have had a brain roughly comparable in size to those of early humans like *Homo erectus*.

Russell's ideas became notorious, but not strictly speaking because he suggested that non-bird dinosaurs could have become smart. Instead, the controversy concerned the way Russell imagined *Troodon*'s anatomy might have evolved to accommodate a large brain. He speculated that *Troodon* would have developed a more vertical posture to support its ever-growing skull. As such it would no longer need a long tail for balance – this feature gradually shrank and disappeared in Russell's thought experiment. *Troodon*'s hands, meanwhile, which had rudimentary opposable thumbs even before the asteroid hit, would have become even better at manipulating small objects. In other words, *Troodon*'s superintelligent descendants – Russell dubbed them 'dinosauroids' – might have looked uncannily human-like.

It's easy to imagine Russell's dinosauroids developing a society and culture like our own: nuclear power, computers, space travel and more all seem within their reach. But that's because dinosauroids basically *are* humans, notwithstanding their dinosaur skin. Other researchers have poured cold water on Russell's thought experiment, which effectively suggests that intelligent animals naturally converge on a human-like body shape.

'It is very unlikely that intelligent dinosaurs would be particularly humanoid,' says Thomas Holtz at the University of Maryland. 'We have really odd anatomies, and there is no reason to think that getting a big brain requires a humanoid body shape.'

Darren Naish at the University of Southampton, UK, has mused on the subject of smart dinosaurs more than most. He thinks there is little fossil evidence to suggest any non-bird dinosaurs really were on the path to human-like intelligence. But if one or more lineages did ultimately evolve to be super-smart, Naish thinks they would probably have retained the features of their prehistoric ancestors: bodies held horizontally with a long tail and a rich covering of feathers. 'Also worth noting is that dinosaurs of course *did* evolve primate-like intelligence,' says Naish. 'Parrots and crows are at that level.'

This makes it more challenging for us to imagine smart dinosaurs making the same technological breakthroughs our species has made. When it comes to envisioning the technological prowess of animals that combine human-like intelligence with the body of a fierce predatory dinosaur, all predictions are off. But it's a safe bet that somewhere in the multiverse, an intelligent dinosaur is wondering what the world would look like if small, furry mammals ruled the Earth.

What would Earth be like without us?

? What was the planet like before *Homo sapiens*, and would it still be that way if we had never gone global? Christopher Kemp rewinds time, erases our ancestors, and hits play.

Imagine for a moment that the last 125,000 years of Earth's history exist somewhere on a tape – a thick, old-fashioned ribbon loaded between two metal drums. With every second that passes, more tape slowly unspools from one drum and is wound onto the other. Now suppose it's possible to stop the tape, to intercede, and to reverse its direction. Rewind. Gradually, with each turn of the drum, our existence is removed.

Every minute, an area of natural forest and woodland the size of ten football fields is restored. At first, for each year that is regained an area slightly larger than Denmark is reforested. It takes only about 150 years of this to restore most of what has been lost. At the same time, urban sprawl retreats like a concrete tide. Megacities shrink to cities and then dwindle into towns and villages, green swathes of pristine undeveloped land reappearing in their wake. The world's rivers are undammed. The sea floor is cleared of its wrecks and its tangled cables. The ozone layer is restored. The remains of most of the estimated 108 billion people who have ever lived are removed from the ground, and fossil fuels, precious stones and metals, and other mined materials are put back in. Tonnes of pollutants, including carbon and sulphur dioxide, are sucked out of the atmosphere.

Finally, we arrive at a point that seems incredibly distant to us: 125,000 years ago. In geological terms it might as well be yesterday, but the span of time between then and now represents the entirety of modern human existence. By running the tape backward to this point, we have removed almost all human impact on Earth. What is it like?

A hundred and twenty-five thousand years ago, Earth was partway through the Eemian interglacial period – a 15,000-year-long temperate phase bookended by two much longer, colder glacials. Suddenly, it had become a warm and green world. In the northern hemisphere, continental ice sheets had retreated from as far south as Germany in Europe and Illinois in North America. 'It got a little bit warmer than at present, and sea levels were maybe a little bit higher at their maximum,' says Ian Tattersall, curator of anthropology at the American Museum of Natural History in New York City.

One of the beneficiaries of this warm and stable climate was *Homo sapiens*. Our species had first appeared around 200,000 years ago in east Africa. By 125,000 years ago the population was probably somewhere between 10,000 and 100,000, surviving by foraging and hunting and making its first forays out of its ancestral home. But we were not alone. 'There were at least three lineages of hominids around,' says Tattersall, an expert in early human evolution. 'There was *Homo sapiens* in Africa; there was the lineage of *Homo erectus* in eastern Asia, which later became extinct; and there were the Neanderthals in Europe.'

Other human species too, both unknown and partly known to us, were struggling to survive elsewhere. 'Who knows what was going on in Africa?' says Tattersall. 'There were hominids in Africa that didn't look exactly like a modern *Homo sapiens*.' The world also would have been teeming with large animals – whales in the ocean, giant herds of herbivores on land. 'I think if you could just teleport into this world, the thing you'd notice right away would be the megafauna,' says environmental historian Jed Kaplan at the University of Geneva's Institute for Environmental Sciences in Switzerland. 'You would find all of these massive herds of big animals roaming around all over the world,' he says. 'There would be woolly mammoths roaming the Arctic. For sure you would see things like bison. You would have big cats living in Europe, maybe horses in the Americas, certainly many more bears, wolves and all of these kinds of herd animals.'

But then, without warning, everything changed. Or more precisely, humans changed first, and then so did the world. 'The shit really didn't hit the fan until humans started behaving in a modern fashion, about 100,000 years ago,' Tattersall says. 'And it was after this that humans sort of stepped outside nature and found themselves in opposition to it, and started all the shenanigans that we're familiar with today.'

It is sobering to read even an incomplete list of the shenanigans that Tattersall is talking about. As recently as about 2000 BC, world population was counted in the tens of millions. By AD 1700, it was at about 600 million; it is now slightly more than 7 billion and grows by an estimated 220,000 people every day. And that's just the humans. According to the United Nations Food and Agriculture Organization (FAO), the global cattle population is 1.4 billion, there are roughly a billion pigs and sheep, and 19 billion chickens worldwide at any one time, almost three for every person.

As befits our numbers, we consume energy like never before. In the twentieth century alone, energy use grew sixteen-fold. According to an article published in 2009 in the *International Journal of Oil, Gas and Coal Technology*, since 1870 an estimated 944 billion barrels – or 135 billion tonnes – of oil have been extracted from beneath Earth's surface. In 2011 alone, the US mined more than a billion tonnes of coal, and China three times as much.

We have also altered the landscape in untold ways. Together, agriculture and the use of fire have tamed and shaped the environment almost everywhere. In many regions, farmed land has replaced the natural vegetation. Between 30 and 50 per cent of the planet's land surface is used in one way or another by humans, and we are tapping more than half of the world's accessible fresh water.

Rice production, in particular, has flattened entire ecosystems. 'People produce little dams,' says Erle Ellis, an environmental scientist at the University of Maryland. 'And that changes the whole sediment movement in a watershed. The goal is to create wetlands everywhere to grow rice. And that has flattened a lot of places. It's impressive.'

In the modern world, we are left with very few places that look the way they would if humans had not intervened.

'There's very few landscapes that are really left, especially in Europe,' says Kaplan. 'There are hardly any forests where you find big dead trees just laying down on the floor. It's incredibly rare.'

Ever since modern humans began to oppose the rest of nature, they moved, dispersing across the world like seeds in the wind, settling in the Near East 125,000 years ago, South Asia 50,000 years ago, Europe 43,000 years ago, Australia 40,000 years ago and the Americas between 30,000 and 15,000 years ago. The final significant, habitable land mass to be settled was New Zealand about 700 years ago.

Everywhere they went, humans took animals with them, some deliberately (dogs, cats, pigs) and others by accident (rats). The introduction of a non-native species to a delicately balanced ecosystem can have irreversible effects, says Ellis. Especially rats. 'They have a huge effect. Anything that nests on the ground or in any place where a rat can get to it – those species are toast.'

We are also efficient killers in our own right, of course. Many species are known to have been hunted or persecuted to oblivion, most famously the dodo (last confirmed sighting in 1662). Also gone: Steller's sea cow (1768); the bluebuck (1800); the Mauritius blue pigeon (1826); the great auk (1852); the sea mink (1860); the Falkland Islands wolf (1876); the passenger pigeon (1914) and the Caribbean monk seal (1952). Many more species have disappeared on our watch. The human march across the globe was followed by wave after wave of megafauna extinctions. The causes are still debated, but many point the finger at us. 'I really think that humans had a role in tipping a lot of these megafauna populations toward extinction,' says Kaplan.

Fifteen thousand years ago, for example, humans were

entering North America from Siberia. 'There was an unprecedented pulse of extinction,' says Bill Ruddiman, a climate scientist at the University of Virginia. 'That requires something brand new, and humans were brand new.'

'The American west, the plains, had a variety that was far richer than the Serengeti today,' says Ruddiman. 'It was an amazing place. Aside from mammoths and mastodons, there were sabre-toothed tigers, horses, camels, gigantic ground sloths – all kinds of animals that went extinct in a pretty brief interval. The best data on that suggests it happened about 15,000 years ago.' Today, wide open – and mostly empty – the American West looks vastly different from the way it did 125,000 years ago.

Removal of large animal species by humans has had effects on the landscape that are apparent almost everywhere. 'A lot of land would be semi-open, kept partly open by these big herds of grazers and browsers and predators,' says Kaplan. 'It's important to keep in mind that landscape is also shaped by animals. These giant herds of bison would be trampling down little trees and keeping the landscape open, certainly not as much as people who are using fire, but definitely having an effect.'

We have also emptied the oceans. According to a 2010 report, the UK's fishing fleet works seventeen times harder than it did in the 1880s to net the same amount of fish. The FAO estimates that more than half of world's coastal fisheries are over-exploited.

Whaling has also changed the oceans beyond recognition. During the twentieth century, several species were hunted to the brink of extinction, and populations have still not recovered. A controversial study published in *Science* claimed that pre-whaling populations were dramatically higher than previously thought. By this estimate, there were once 1.5 million humpback whales, rather than the 100,000 estimated by the International Whaling Commission. It is a similar story for minke, bowhead and sperm whales.

We have also shifted the climate. In 2013, atmospheric carbon dioxide levels topped 400 parts per million for the first time in millions of years; 125,000 years ago they were 275 parts per million. The increase comes partly from the burning of fossil fuels but also from the stripping of the world's forests, which have acted as an almost bottomless carbon sink for millions of years.

The impact is etched dramatically on Earth's ice. Across the world, glaciers are retreating and in some places have disappeared. The US National Snow and Ice Data Center at the University of Colorado in Boulder maintains an inventory of more than 130,000 glaciers around the world. Some are growing; many more are shrinking. Worldwide, for every glacier that is advancing, at least ten are retreating. At its creation in 1910, Glacier National Park in Montana had an estimated 150 glaciers. Today there are about thirty, all of which have shrunk. In 2009, the Chacaltaya glacier in Bolivia – once the location of the world's highest ski-lift – disappeared. The polar ice sheets are breaking apart, calving city-size blocks of ice into the oceans. In 2013, a 30-kilometre crack in the Pine Island glacier in Antarctica created an iceberg the size of New York.

By running the tape of time backwards, almost all of these human impacts on Earth are gone. Now, just for fun, let's do something else: let's remove *Homo sapiens*. Imagine that 125,000 years ago, our small band of ancestors in east Africa was wiped out by a catastrophe: a lethal virus, perhaps, or a natural disaster.

Now, let the tape run forward again. What would the world look like today if modern humans had never been here? In some respects the answer is obvious: it would look a lot like the world of 125,000 years ago. 'We would have a continuous biosphere – one that we can scarcely now imagine. That is, forest, savannahs and suchlike, extending across the Earth,' says Jan Zalasiewicz, a geologist at the University of Leicester, UK. 'No roads. No fields. No towns. Nothing like that.' The land would teem with large animals, the seas with whales and fish.

But it wouldn't last, says Ruddiman. If humans had died out 125,000 years ago, we would now be entering another ice age. Glaciers would be growing and advancing. It's a controversial idea and it has earned Ruddiman his critics. But now, more than a decade since he first proposed it, many climate scientists agree with him.

'If you erase the human effect there would be considerably more sea ice and much more extensive tundra around the Arctic circle,' he says. 'Boreal forest would have retreated and, most dramatically of all, you would have growing ice sheets in a number of northern regions – the northern Rockies, the Canadian archipelago, parts of northern Siberia. It's the very early stages of an ice age. That's the single most dramatic change.'

Or maybe not. Perhaps, in our absence, one of the other human species that was present – Neanderthals, *Homo erectus*, or an as-yet unidentified species – rises to prominence and begins to shape the world instead of us. Tattersall is doubtful. 'Having established themselves, would they have followed in our footsteps?' he says. 'Would they have become an ersatz *Homo sapiens*, implying that there was some sort of inevitability on our having become what we became? I would guess no.'

But there is a delicious counterpoint to this argument.

'There is this idea – convergent evolution – that if we didn't come along and do all this, somebody else would,' says

David Grinspoon, curator of astrobiology at the Denver Museum of Nature and Science in Colorado. 'There still would have been selective pressure for some other species to go through the same kind of development that we did, where there's this feedback between big brains, and language, and symbolic thought, and developing agriculture. If the scenario is literally that just *Homo sapiens* goes extinct but it's still the same general landscape, maybe something similar would have happened. It wouldn't have been identical because there's so much randomness, and it might have taken longer.'

In short, perhaps it all would have happened anyway. Maybe this modern version of Earth, and our place in it, was unavoidable. Remove *Homo sapiens* from the equation, reforest the world and repopulate it with megafauna, and maybe in 100,000 years or so our greatest works, our advancements and our errors – or at least something like them – would still be the outcome. 'I wish I had a crystal ball, or an alternate-universe viewer,' says Grinspoon. 'It would be great to know.'

Could electric motors have powered the industrial revolution?

? The village church, the village green – the village power mill? What if electric motors had pre-empted the age of steam, wonders A. Bowdoin Van Riper.

Eighteenth-century scientists thought of electricity and magnetism as substances, 'imponderable fluids' whose particles were too small and subtle to be detected by ordinary instruments. In their eyes, the two fluids were utterly separate and distinct. It was no more possible to transform electricity into magnetism than turn water into wine (without divine assistance, anyway).

English chemist and physicist Michael Faraday saw it differently. Early in the 1820s, Hans Christian Ørsted and André-Marie Ampère had shown that an electric current moving through a wire generated a magnetic field around the wire. Building on their work, Faraday showed in 1831 that the reverse was also true: moving a wire through a magnetic field creates an electric current in the wire. It was Faraday who drew the revolutionary conclusion that electricity and magnetism were two manifestations of a single phenomenon, and it was also Faraday who recognised the technological implications. The use of an electric current to generate a magnetic field became the basis of the electric motor, and the use of magnetic fields to create an electric current became the basis of the electric generator.

Faraday's conceptual breakthrough happened when it did for identifiable reasons. One was the invention of the battery, which could provide a steady flow of electricity. Another was the Romantic movement, which promoted a holistic view of the world that encouraged scientists to consider that seemingly discrete phenomena might be connected. Suppose, however, that a similar set of causes had come together a century earlier. What if some Enlightenment-era Faraday in a powdered wig had made the crucial breakthrough? What if electric generators and motors had been on hand before the industrial revolution began?

In this alternate timeline, the first electric motors would probably have arrived on the market some time in the 1740s – a time when the still-new steam engine was used only in a few niche applications like mine drainage. Potential users would, therefore, have judged electric motors not against steam engines but against the power sources that had served for centuries: wind and water for heavy-duty work, human and animal muscle for everything else. Especially when judged against muscle power, electric motors would offer obvious advantages – compactness, quiet operation and the ability to work steadily for hours with no need for food, water or rest. In the 1740s and for decades afterwards, steam engines had none of those virtues and were more expensive to boot. Electric motors would, therefore, have been adopted more widely and more quickly than steam.

The electric motors of the 1740s would have been small and quiet enough to operate in a modest-sized workshop. Opportunities to apply them there would have abounded. In the textile industry alone, their rotating shafts could have driven spinning wheels, yarn winders and knitting machines. Elsewhere, they could have powered blacksmiths' blowers, cabinet-makers' drills, potters' wheels or rope-makers' hemp-twisting cranks. The means to drive generators to power the motors was already at hand: waterwheels or windmills. The principal difference between a 'generating mill', designed to turn a light electricity generator's shaft rapidly, and a traditional grinding mill, designed to turn a heavy stone slowly, would have involved new gear ratios that, once worked out, could have been replicated easily by any experienced millwright. Once operating, a single generating mill could have served many customers who located, or relocated, their workshops near by.

The steam-driven industrial revolution that actually took place in the final third of the eighteenth century emphasised centralisation. Even the most sophisticated steam engines of the time were so large, expensive and fuel-hungry that to use them efficiently, you needed large factories. The electrically driven industrial revolution that might have taken place in the late eighteenth century would, at least at first, have been inherently decentralised. Small, inexpensive electric motors could have been readily integrated into existing workshops.

Larger factories would doubtless have followed, but they would have been an option rather than a technological and economic necessity.

An industrial revolution rooted in electric power rather than steam would also have had effects beyond the workshop. In our world, the electricity distribution system developed after, and in imitation of, the distribution system for natural gas. Electric power was produced at large centralised facilities and distributed to homes and businesses through a branching network of wires. However, had electricity come into widespread use in the mid-eighteenth century, there would have been no such model to emulate.

The provision of electricity might have been organised more like the provision of heat, with those in the countryside opting for self-sufficiency (a waterwheel, windmill or small boiler-and-turbine unit to run a generator, say) while those in cities could have chosen their supplier from one of several competing neighbourhood sources, as they did for coal deliveries. Small, localised power grids might have become the rule, and large, city-spanning ones the rare exception.

The longer-term effects of an electrified industrial revolution would have been profound. The lateral sprawl of the world's major cities and the rise of the suburb – driven by the electric subway and tram – would have begun far earlier and possibly progressed further than in our world. The spread of electric lighting and, in its wake, electrical home appliances would also have begun earlier. The electric automobile as personal transportation would have prospered and maybe even evolved into a mature technology before the invention of the internal combustion engine. Gas lamps and gas stoves might well have been stillborn – why build an expensive and potentially explosive system of gasworks and mains if existing electrical systems could do the job? Cities built without networks of gas pipes would have been less prone to burn in the event of catastrophic damage, as San Francisco did after the 1906 earthquake. The absence of a city-wide electrical grid would, in turn, make massive power outages a virtual impossibility.

Would it all have produced a better world? Perhaps not, but it is tantalising to contemplate an industrial revolution whose hallmarks were not smoke, grime and the hiss of steam power but the quiet whirr of electric motors and the glow of pure, bright light.

What would a world without fossil fuels look like?

? Our advanced civilisation is built on easily exploited coal, oil and gas. Michael Le Page explores an alternative history that holds lessons for us all.

You are probably reading this on a piece of ex-tree. Felled by a petrol-guzzling chainsaw, it was carted to a paper mill in a diesel-powered truck. Or perhaps these sentences are on a tablet, with plastic components that started life as crude oil, and metal smelted with coke produced from the tar sands of Canada. Either way, the words are probably lit with electricity from a coal-fired power station. Maybe you are even sipping wine, from grapes grown with fertiliser made using natural gas, in a glass created in an oil-fired furnace.

The list goes on and on. Our civilisation is built on fossil fuels. We depend on them not just for energy but for all kinds of raw materials and even the food we eat. Weaning ourselves off this stuff is not going to be easy.

But what if there were no fossil fuels on Earth, or at least none that were easy to exploit? Would history have taken a different course? What of the industrial revolution? Would modern civilisation even exist? To tackle such questions, we must go on a journey that will carry us from medieval times into a counterfactual yet strangely familiar world. It's a fascinating thought experiment with implications for the future of our own planet, as well as for alien civilisations.

A long time ago we weren't too bothered about fossil fuels. Early civilisations sometimes used them if they were there for the taking, but did not rely on them. Wood or charcoal were superior for just about all purposes. In medieval Europe, only Britain began to use coal on a large scale. Why? The answer has to do with geography, argues historian Rolf Sieferle of the University of St Gallen in Switzerland.

In the thirteenth century, water transport was cheap. Floating wood down a river required almost no energy, and a single horse could pull a barge weighing 50 tonnes. Going overland was far more difficult: a horse could pull only 1 or 2 tonnes. If wood had to be hauled any distance, its price rose rapidly. This was bad news for Britain, which lacked the long navigable rivers of continental Europe. What it did have, though, were large outcrops of coal right next to the River Tyne in northern England, which could be loaded on to ships and taken by river and sea south to London.

Coal was regarded as a dirty, stinky fuel that wrecked people's health. But in Britain it became cheaper than wood, and caught on because it could replace wood for many purposes – making lime for mortar, firing bricks, heating homes and so on. Yet raw coal is inferior to wood in one crucial way – it is no use for smelting copper or iron ores. The impurities in coal weaken the metal. Instead, smelters relied on charcoal, says Benjamin Roberts of Durham University in the UK, who studies early metallurgy.

Then, in the seventeenth century, English inventor Hugh

Plat suggested that coal could be purified by 'charring' it – a process called coking. In 1642, brewers began to use coke for malting, achieving a new lighter roast that led to the production of the first pale ale, and in 1709, Abraham Darby started using coke to smelt iron ore, ending the dependence on charcoal. This, some argue, was the pivotal innovation that kick-started the industrial revolution in the eighteenth century, by providing a plentiful supply of cheap iron which led to many other innovations such as iron railways.

The question is, could wood alone have fired the industrial revolution?

Before the eighteenth century, the main source of energy was the solar power captured by plants – biomass. The energy needed to pull ploughs or haul loads came from the food consumed by animals or labourers. The heat for cooking, heating and industry came mainly from wood. Plants were also the ultimate source of many raw materials, from wool and cotton to timber for houses and ships. Other materials, from cement to iron, couldn't be made without burning wood or charcoal. This meant that as populations grew and energy use rose, land area became a limiting factor. Negative feedbacks kicked in. If charcoal burners used more wood, there was less for building and ship-making. If more trees were planted, there was less land for growing food.

Plenty of things could be done to relieve these constraints, from better growing practices such as coppicing to trading with neighbouring countries or taking land by force. But as growth continued, the limits were soon reached again. 'Expansion in one area meant contraction elsewhere,' says historian Tony Wrigley of the University of Cambridge.

Exploiting coal, however, gave us access to a virtual forest bigger than all the continents put together – millions and millions of years of accumulated plant growth. It's a finite resource, but one so vast we still haven't exhausted it.

Coal enabled Britain to produce iron cheaply, without large forests. Iron farm tools boosted productivity and iron railways made it cheaper to move that food and other goods. 'There were positive feedbacks,' says Wrigley. 'Each advance made the next more likely.'

Here's how important coal was: by the 1820s, growing enough wood to replace all the coal used in Britain would have required a land area larger than the country. 'It is inconceivable that there would have been a development worthy of the name industrial revolution based on charcoal,' Sieferle writes in his book *The Subterranean Forest: Energy systems and the industrial revolution.*

What if we had no coal or other fossil fuels? By the nineteenth century, says Sieferle, the growth constraints imposed by limited land area would have kicked in. In earlier civilisations, periods of rapid advance were often followed by collapse, he says. 'There was no collapse in Europe because we had fossil fuels.'

Clearly, a world without fossil fuels could not develop far without an energy source that does not depend on biomass. There are many other potential energy sources – the key question is whether any can be exploited without advanced technology.

For a start, we can rule out nuclear power and solar photovoltaics. By contrast, solar thermal power is easily exploited to heat air or water using little more than black pipes. An array of mirrors can focus enough heat to melt metal. But even today exploiting this on an industrial scale is a huge challenge. In a world without fossil fuels, the sun might be used for cooking or heating homes and water, but as an energy source for industry it's probably not viable.

That leaves wind and water. On the sea, wind powered

most trade and exploration until well into the nineteenth century. On land, wind and water mills have been used for at least two thousand years. In medieval Europe the technology reached new heights. By around 1600, wind and water mills were grinding grain, sawing wood, boring pipes, polishing glass, drilling holes, pressing seeds for oil, grinding up stones, pumping out mines and more. It was in these mills, some argue, that the foundations for the industrial revolution were laid.

Water continued to power factories throughout the industrial revolution. It wasn't until 1820 that coal-powered steam engines overtook water as the main source of mechanical power in British factories, says Terry Reynolds of Michigan Technological University in Houghton, and author of *Stronger than a Hundred Men: A history of the vertical water wheel.* 'In the US, the date was around 1870.'

Factories turned to coal in part because they were using pretty much all of the available water power already. Most factories had to be located near ports and navigable rivers, where hydropower resources are limited. 'In Britain in the late eighteenth century some streams were so dammed up that no further power could be squeezed out of them,' Reynolds says. There were vast untapped hydropower resources in distant hills and mountains, but there was no way to get the energy to where it was needed.

In our world, two technologies changed this. First, pioneers like Michael Faraday worked out how to transform motion into electrical power. Later in the nineteenth century, engineers worked out how to transmit electricity over long distances. 'Electric power transmission transformed hydropower into a vital contributor to the modern world's power mix,' says Reynolds.

It seems clear, then, that to develop on the basis of

hydropower rather than fossil fuels, a civilisation would have to develop hydroelectricity. Is that feasible? Without fossil fuels, western Europe could have developed technologically to around the 1800 level, Sieferle thinks. This was the time when innovators like Faraday and Humphrey Davy were developing the basic components of modern electrical technology.

So where would this alternative path take us? Forget steampunk: this would be a hydroelectropunk world. For a start, industrialisation might have taken off in the mountains of Norway or Switzerland rather than Britain. Without easily exploitable fossil fuels, economic and technological development, and population growth, would certainly have been slower. Regions with abundant hydropower resources, including Scandinavia, Canada and parts of South America and Africa, would have had a huge advantage.

Without dirty coal fires, cities would not be grimy and smoke-stained. Driven by the need for better hydroelectric generators rather than more efficient steam engines, science would focus more on electrodynamics than thermodynamics, perhaps hastening the development of smaller, lighter electric motors and batteries. The pace of life would also be slower – going from London to New York would involve weeks spent on an iron sailing ship rather than hours on an aeroplane. The world might even be more equitable: without steam power, some European nations may have struggled to establish vast empires. Most important of all, there would be no impending climate doom.

It is an appealing vision, but perhaps an unlikely one. We tend to assume that technological progress is inevitable, but it may not be. China started smelting iron with coke as early as the ninth century, for instance, but the industrial revolution did not happen there. It happened in Britain only because a very particular set of geographic, social, economic and cultural factors came together.

Hydroelectricity is a more complex technology than steam power, and even if developed by a pre-industrial civilisation there are reasons to doubt that it could drive an industrial revolution. With relatively simple technology, electricity can be used for lighting, heating, cooking, powering engines, making fertiliser and even melting metals. However, while electric power can reduce the amount of charcoal needed for smelting iron by more than half, it cannot replace it entirely. So wood would still be a limiting factor, and without cheap iron to make tools, machines and railways, rapid industrialisation would be extremely difficult.

Scaling up hydroelectricity is also more challenging than scaling up coal mining. It was possible to build dams long before the industrial age. In 1177, for example, the 400-metre-long Bazacle dam in Toulouse, France, was built to power water mills. In 1888, it was converted to a 4-megawatt hydro-electric power station.

However, it would take 2,500 dams of this size to provide the same amount of energy as England and Wales extracted from coal at the beginning of the nineteenth century. To match coal consumption in 1850, it would take another 8,000 dams. And what do you construct them with, asks Sieferle. Building materials like cement and bricks became cheap in our world only because we could use coal to produce lime for cement and to fire bricks. Then there's the energy needed to transport materials. Building hydroelectric dams takes a lot of planning and labour and energy, and yet the dams don't provide energy until they are finished. The economics might not add up.

So could civilisations make the transition to advanced

industrial economies without exploiting fossil fuels? 'You cannot exclude this, of course,' says Sieferle, but he doesn't think it likely. Historian Kenneth Pomeranz, author of the *The Great Divergence: China, Europe, and the making of the modern world economy*, is more optimistic. 'My guess would be slower growth,' he says. 'But there are obviously a lot of unknowns.' And slower growth is not without its dangers. Regional civilisations are vulnerable to epidemics, famine, earthquakes, war and volcanic eruptions – the downfall of many. The longer it takes for a global industrial society to emerge, the more chance that disaster will strike.

All this hints at a universal truth: wherever technologically advanced civilisations develop, most – perhaps all – do so by exploiting fossil fuels. 'You can't change anything without expending energy,' says Wrigley. 'Energy is an absolute prerequisite.'

Fossil fuels probably exist on most worlds where carbon-based life might thrive. Perhaps there are aquatic alien civilisations that have progressed technologically without exploiting fire, or even some that flourish on worlds with no oxygen in the atmosphere, but it's hard to imagine how this could happen. If fossil fuels are key to developing advanced technology, global warming may not be a uniquely human problem but an issue for petrol-heads across the universe. The laws of physics apply everywhere, after all. 'The downside of access to a stock of energy in fossil fuels is very great,' says Wrigley. 'And making the transition to other energy sources is far from easy.'

How many alien civilisations have wrecked their planet or run out of fossil fuels – or both – before they managed to make the transition to sustainable energy sources? This could even be the 'great filter', says James Kasting of Pennsylvania State University, who studies what makes planets habitable.

Kasting refers to the idea that our failure to discover intelligent life elsewhere despite the vast number of stars – the Fermi paradox – is due to some bottleneck that either stops intelligent life evolving, or makes it go extinct.

After such a collapse occurs, there would be no easily exploitable fossil fuels left. There might a brief window of opportunity for a civilisation to re-establish itself using more sustainable energy sources, before all knowledge and resources had been lost. But after that, it would be too late. If this is right, each planet essentially offers one chance to make the transition to an advanced and sustainable state. We had better not waste ours.

What if we could start over? Skip to page 106 to discover Civilisation 2.0.

What if the Enlightenment had spluttered out?

? Science in England was on a roll once the 'merry monarch', Charles II, began backing experiments – and the Royal Society was born. But what if the king had rejected empirical science? asks William Lynch.

Newton, Einstein, Darwin . . . yes, they're all worthy of a 'what if?' But counterfactual history isn't just about the celebrities of scientific history. One unlikely figure who also made a difference was Charles II, king of England, Scotland and Ireland from 1660 to 1685. Were it not for Charles's enthusiasm for a particular branch of natural philosophy, modern science might have turned out very differently.

On 28 November 1660, twelve men met in Gresham College, London, to found what would become the Royal Society of

London. At only their second meeting, the group found sponsorship from Charles II, newly returned from exile to take up the vacant throne.

It was a significant moment. The king's approval led to a royal charter in 1662, giving the society the right to publish under its own name. The Royal Society became the visible vanguard of natural science, through the work of prominent members such as Robert Boyle and Isaac Newton, and through the society's journal, *Philosophical Transactions.*

Whereas continental European figures such as René Descartes and Galileo Galilei promoted a philosophical, deductive approach to science, modelled on that of geometry, the thinkers of the Royal Society were followers of the scholar Francis Bacon's inductive and experimental approach. This experimental emphasis was soon taken up throughout England, and eventually the world.

Without Charles's support for the Royal Society, might science have taken a different, more philosophical and deductive path – and been crippled by its indifference to experiment? Two factors just might have led the king to choose otherwise. Many of the society's members had connections to the regime of Oliver Cromwell, who had executed Charles's father, Charles I, in 1649. And the younger Charles had been tutored while in exile by the philosopher Thomas Hobbes, who was much more in the European intellectual mould.

Indeed, Hobbes launched a polemical war against Boyle and the Royal Society for basically mucking around with machines and calling it philosophy. He thought that rather than focusing on artificial experiments, science should use deductive reasoning to discover how nature ordinarily behaves. So although Boyle's air pump might have appeared to create a vacuum, anyone reasoning correctly should have concluded that some air would find its way inside the machine.

Some historians of science believe that Hobbes's point of view might have prevailed. Steven Shapin of Harvard University and Simon Schaffer of the University of Cambridge have argued that his criticisms of the experimental approach were cogent enough to have steered science in a different direction. In that case, our modern focus on experiments is merely an accident of the political history of Restoration Britain.

Without that accident, Hobbesian science would have been the mainstream. Scientists would have been mainly concerned with whether an explanation is theoretically conceivable, not whether it is testable. They might, for example, have ignored the pneumatic physics of the air pump, which could have prevented the development of steam engines, and also removed a crucial tool of the eighteenth-century chemical revolution – the collection and manipulation of gases. No steam engines and no modern chemistry mean no industrial revolution, and therefore no high-tech modern science such as particle accelerators. We would still be arguing over the existence of atoms.

But this is highly unlikely. The lesson from history is that political approval is not the only force driving experimentalism. Continental European thinkers, for example, found ways around their geometry-centred way of thinking. They gradually bent the rules, slowly learning to refer to individual experiments. For instance, Blaise Pascal conducted experiments with the barometer, but reported them as general 'experiences' rather than concrete experimental data.

While Boyle objected that this seemed to blur the line between real experiments and thought experiments, Newton's own synthesis owes more to this tradition of blended mathematics and experimentalism than to his Royal Society colleagues' emphasis on theory-free 'matters of fact'.

If Charles II had not established the Royal Society, and

had laid down Hobbes's philosophy as the approach to be taught in schools, it would not have stopped even self-styled Hobbesians from tinkering with air pumps. Official Hobbesian science would still have seen plenty of experimental work, such as Newton's optics or public demonstrations of electricity and magnetism, and the scientific puzzles raised would invite further experimentation – even if disguised by a more deductive theoretical framework. And no doubt the Baconians would still have pursued their interests, with or without government patronage.

Most importantly, technology would have continued to advance. Improvements such as the scientific instruments that accompanied Britain's growth as a sea power were not dependent on academic science. Latter-day Hobbesians would have had to stretch their theories to explain such developments. And new tools and ever more skilled craftsmen would have provided increasingly tempting opportunities to muck around with instruments and call it philosophy.

Could anyone but Newton have put the heavens in order?

? A squabble over the colour of light almost exiled the young mathematician from science. Peter Rowlands asks what would have happened if Newton had never returned.

On 6 February 1672, the twenty-nine-year-old Isaac Newton confidently dispatched his first paper to the Royal Society of London – a body which, just a few months earlier, had roundly applauded him for his invention of the reflecting telescope. Newton expected a similar response to his paper's assertion that white light is a mixture of many colours.

But the startling new theory generated a storm of controversy that continued for years afterwards. Newton was totally unprepared for anything other than full acceptance, and the Royal Society's reaction left him aghast.

As the controversy raged on, Newton gradually lost interest in science and finally threw in the towel in 1678. For six years, he resisted the Royal Society's attempts to coax him back. Newton was eventually persuaded to return when the society's big shots insisted that only he had the mathematical skills to find a law of planetary motion.

The outcome was the *Principia*, one of the greatest scientific books ever written. Here Newton explained the motion of planets, tides, the precession of equinoxes and the acceleration of falling objects, all using his new mathematical theory of forces.

So just how different would history be if Newton had not returned to science? Wouldn't others have made the same discoveries? After all, people had already guessed that there must be some sort of law of attraction between the planets, similar to the one Newton found. My answer is no. There was something unique about Newton and the period in which he lived that made a huge difference. If Newton had carried out his threat to give up science altogether, today's world would be very different from the one we know.

During the seventeenth century, scientists such as Robert Hooke were viewed mainly as inventors. Their explanations of natural phenomena were in terms of mechanical devices and models rather than fundamental mathematical principles. Newton changed all that. By realising that there were abstract, universal laws that dictated the behaviour of the physical world, he made the scientist, above all the physicist, a towering authority capable of grasping all the secrets of the universe. Newton's style of thinking was also unique. Only by abandoning familiarity and common sense could he invent abstract laws that covered myriad, seemingly unconnected phenomena and imagine forces, such as gravity, without any conceivable physical cause.

The timing was unique too. Newton was the last in a long line of medieval theologians who believed that everything could be explained by a single underlying idea – a philosophical stance that was becoming seriously passé by the late seventeenth century. Even Newton's most brilliant contemporaries refused to believe that, for example, gravity instantly affected everything in the universe, from planets to particles of matter. How could an apple exert the same force on Earth as Earth exerts on an apple? It defied common sense.

We know what might have happened without Newton because of what did happen before his ideas became widely accepted in the mid-eighteenth century. Before he was converted to Newtonianism, Swiss mathematician Leonhard Euler explained the motion of planets by supposing that whirlpools in a gaseous ether kept them in orbit around the sun. Several years later, Euler abandoned this approach and accepted Newton's abstract law of gravity. He went on to create our modern formulation of Newtonian dynamics.

But without Newton, Euler's original idea might have lived on. The mathematicians Joseph-Louis Lagrange and Pierre-Simon Laplace would have extended it, no doubt, using ever more brilliant maths. Difficulties such as the rotation of galaxies would be solved by invoking specific fixes. That ad hoc style of reasoning would have dominated non-Newtonian science. Few people would even think in terms of universal mathematical laws. Every phenomenon would have a specific explanation, and no one would conceive of putting forward a theory to account for more than the existing data. Generalising would be considered a pointless exercise, not worth funding and systematically denied publication. It would be thought absurd to worry about solving fundamental problems when there were more immediate practicalities to be dealt with. Physicists would not have won their present status as ultimate philosophers.

The effect on technological progress would have been profound. Michael Faraday's discoveries in electricity and magnetism, which led directly to the electric motor and dynamo, would have baffled his mid-nineteenth century contemporaries. James Clerk Maxwell would never have explained Faraday's findings by writing the equations that unify electricity, magnetism and optics in a single theory. Today's scientists would only just be beginning to understand electromagnetism. At best we would have a plethora of specific theories enabling limited technological development in such things as electrical power generation and radio. There would be no computers or any other advanced electronic devices. The technological developments of the second half of the twentieth century simply wouldn't have happened.

And what of quantum mechanics? Anyone suggesting that light is a particle, as Newton proposed later in his life, would be viewed as a crank. And the idea that light can be either a wave or a particle – one of the cornerstones of quantum mechanics – would be inconceivable. Certainly there would be little concept of physics as we know it. Chemists, with their ideas about atoms, molecules and the periodic table, would be considered true scientists. But physicists would be called kinematical engineers and would have a much lower status.

There would be many unexplained phenomena, but there wouldn't be any anomalies in a general theory – because there wouldn't be one. Radioactivity would be chalked up as just another curiosity. And without the sudden and brilliant success of Newton's rather 'unscientific' and essentially metaphysical way of thinking, we wouldn't be looking for a grand unifying theory.

In that search, today's physicists talk about unifying the four forces of nature into one. But I think their quest for the theory of everything repeats many of the mistakes of Newton's contemporaries. Even ideas such as string theory are too anchored in existing knowledge and lack the conceptual originality required. We need another Newton as never before.

What if Darwin had not sailed on the *Beagle*?

? Other people could have had Charles Darwin's insights – even did – but no one else could have got them accepted by the establishment, says John Waller.

In early September 1830, Charles Darwin, just twenty-one years old, nervously approached the Admiralty buildings in London. A week before, he had received a letter inviting him to serve as captain's companion on HMS *Beagle* as it charted the coastal waters of South America. Already a passionate naturalist, his spirits had soared – only to be dashed by his physician father Robert, who said he could not go. Charles was meant to be donning a dog collar and settling down to life as a parson, not gadding around the world bagging wildlife. Luckily, his father was persuaded to relent by his brother-in-law Josiah, who learned about the trip while seeking advice from the physician about a 'buffy discharge from his bowels'.

But even now the way wasn't clear. The *Beagle's* captain, Robert FitzRoy, was subject to depressive episodes and wanted a companion to ease the burdens of command. On paper at least, Darwin was not a good proposition. FitzRoy was a Tory aristocrat – anti-republican, pro-slavery and devoted to the Church of England. Darwin came from a family of Whigs and abolitionists. To make matters worse, as an amateur physiognomist, FitzRoy surmised that Darwin's flabby nose indicated a 'lack of energy and determination'. Luckily, during that first meeting at the Admiralty, Darwin's unaffected enthusiasm won the captain over.

Just how different would history be if uncle Jos had not been troubled by digestive complaints, if Darwin's father had stood his ground, or if FitzRoy had refused to share a cabin with a man so different to him? Would we now see ourselves as fallen angels rather than upright apes? Would creationism rule?

On at least one thing, most Darwin scholars agree: had Darwin not sailed aboard the *Beagle* he would not have arrived at his theory of evolution by natural selection. It took the raw novelty of travel in alien climes to shatter his received views of nature as harmonious, benign and static. So many aspects of wild nature posed awkward, even embarrassing questions for this cosseted young man. And so, in the months after his return, he began questioning the unquestionable: the immutability of species. Isaac Newton had shown that God was a lawmaker, not a tinkerer. So didn't it make more sense to imagine that the different species of Galapagos mockingbirds and South American rheas had arisen as the result of natural laws rather than as the creations of an interventionist deity? As for those fossilised mega-mammals he had dug up in Patagonia – extinct llamas, sloths and armadillos – surely they were the forebears of modern species?

To Darwin in early 1837 it seemed positively medieval to imagine God stepping in with each new geological or climatic shift to create a new suite of species. But the only clear alternative to this orthodox view was politically, religiously and scientifically subversive. The theory of evolution – that organisms change by degrees over time – had been around for decades, but it was an idea that few respectable naturalists would entertain. Yet, driven by what he had seen on the *Beagle* voyage, Darwin secretly became an evolutionist.

He then began the search for a plausible mechanism of change. At first, he favoured an idea expounded by his grandfather, Erasmus, and Jean-Baptiste Lamarck, that organisms progressively adapt to their environments by inheriting characteristics acquired by their parents over the course of their lives. But in the early winter of 1838, Darwin came up with his own theory: natural selection.

It is very hard to imagine Darwin making the same cognitive leaps if he were a vicar in rural England. It was the *Beagle* experience that pushed him towards the heterodox doctrine of evolutionism. But does this really matter historically? After all, Darwin was not the only person of his generation to suggest the idea of natural selection. In 1858, a younger naturalist, Alfred Russel Wallace, suffering from a severe bout of malaria on a small island in the Malay archipelago, arrived at pretty much the same theory. Perhaps, then, had Charles Darwin not sailed aboard the *Beagle*, we would simply refer to 'Wallaceism' instead.

It is important to note here that neither Wallace nor Darwin achieved fame in their lifetimes for coming up with a plausible mechanism for evolution. Until early in the twentieth century, only a minority of biologists accepted natural selection as the main driving force of evolutionary change. Most preferred to invoke some variant of Lamarck's idea of inheriting acquired characteristics. Darwin's great accomplishment in his own time was to persuade other biologists to embrace the idea of evolution itself. And in this he was remarkably successful. Within a decade of the publication of *On the Origin of Species*, roughly 60 per cent of British biologists were card-carrying evolutionists. Wallace contributed to this astonishing achievement, but could he have won through on his own?

Possibly, but it would have been a hard fight. For a start, in 1858 Wallace had only a tiny fraction of the data available to Darwin. For thirty years, Darwin had amassed staggering quantities of evidence on biogeography, taxonomy, embryology, palaeontology, physiology, heredity, anthropology, animal husbandry and horticulture, with which he could back up his argument. Just as importantly, Darwin's previous publications – not least his study of barnacles – had already won him a solid scientific reputation. Wallace had yet to win his scientific spurs.

There were also less creditable reasons why Wallace's campaign might have foundered. In a society riven by class distinctions, he had the misfortune to be the son of a hard-up lawyer. True, the elites of English science did tolerate the 'lower' orders getting involved in science. But since Wallace paid his way by sending rare finds to wealthy collectors back home, his work carried the stigma of trade. And scientific tradesmen, no matter how skilled, were meant to get on with 'fact-gathering' and leave theorising to their social 'betters'.

Wallace's humble origins might not have mattered had he been drawn to a less controversial area of scientific enquiry. But since the late eighteenth century, evolution had been a dangerous idea. Challenging *Genesis* alarmed the religious and political establishments, both of which freely invoked God in defence of monarchy, aristocracy and inequality. Hence evolutionism's popularity among freethinking democrats and republicans. As a result, however, advocates of evolution who, like Wallace, were of humble birth could expect an angry reaction from the scientific elites, many of them clerics and nearly all well heeled.

Darwin was different. By 1859, he was the absolute embodiment of intellectual and social respectability. By backing evolutionism, he wrested it from the grasp of political radicals and made it an acceptable way of looking at the natural world; he made talk of ape-ancestry a fit subject for polite scientific enquiry. It is hard to imagine anyone else in the front ranks of Victorian science who could have conferred the same respectability on the subject. Nor were any of the other big guns keen to try. As evolution's inside man, Darwin made a genuine difference. Only someone of his eminence could have said that humans are mere beasts and finished up with a state funeral in Westminster Abbey.

Without Darwin, we would still – mostly – believe in evolution by natural selection. To think otherwise would be to ignore the huge advances in biological science in the first half of the twentieth century that made the theory irresistible. But although we can't say how long it would have taken biologists to embrace evolutionism without him, one thing is fairly certain: in 1859, Charles Darwin gave the still-fragile theory of evolution the protection it needed to take root and become perhaps the most powerful idea of modern science.

What if Einstein had been ignored?

? Two centuries after Newton fixed the motion of the stars in the heavens, another young man set them askew. But it took a cosmic coincidence for his ideas to be published, write Graham Lawton and Gerald Holton.

Accidents change history, and in the history of science there is surely no greater accident than Max Planck being one of the editors of the German journal *Annalen der Physik* 100 years ago. Without Planck's influence, it is inconceivable that the journal would have published, in the space of a few months, four papers by an unknown amateur scientist – papers that transformed our understanding of the universe.

The unknown was Albert Einstein, then a twenty-six-year-old patent examiner living in Berne, Switzerland. In 1905, Einstein had something of a purple patch. The first three papers he wrote that year – on the photoelectric effect, special relativity and Brownian motion – are now each considered worthy of a Nobel prize in their own right. The fourth laid the groundwork for the famous equation $E=mc^2$.

Planck had never heard of Einstein, who had no formal scientific training beyond a qualification to teach high-school physics. But he saw something in the papers and, having dispatched an assistant to Berne to check that Einstein really existed, decided that they should be published. Planck's support was critical: no other physicist at the time would have accepted the papers, and for years afterwards Planck was widely blamed for allowing such 'unnecessary' work to see the light of day.

The attacks started almost straight away. In 1906, for example, an experimental physicist called Walter Kaufmann at Göttingen University in Germany published the first experimental test of Einstein's paper on special relativity. At the end he wrote: 'I disprove herein Einstein.' Planck remained unmoved, and Einstein himself paid no attention. When his friends pointed out that his career was at stake, he confidently asserted that Kaufmann's experiment must be flawed. (It was – his equipment had sprung a leak, something that was only realised later.) Planck and Einstein were, of course, eventually vindicated.

So what would have happened had Einstein's papers not been accepted, or Kaufmann's attack struck home? When Einstein himself was asked that question later in his life, he replied that he firmly believed another physicist would have come up with the same concepts. He even had an individual in mind: Paul Langevin of the Collège de France in Paris, who is probably best remembered for the scandalous affair he had with Marie Curie in 1910. Langevin studied under mathematician Henri Poincaré, who published an unsuccessful attempt at a theory of relativity in 1905 shortly before Einstein did.

And Einstein had good reason to believe that Langevin had what it took. The two worked together and became close friends; they can be seen standing next to each other in a photograph of the first Solvay conference of eminent scientists in 1911. It's curious to think that the Frenchman's face, and not the one to his immediate right, could have become one of the icons of science. We can never know if Langevin really would have come up with special relativity. But arguably no one but Einstein could have come up with general relativity, the 1915 theory that incorporated gravity into special relativity and gave birth to the science we now call cosmology.

And had Einstein not risen to global fame there is another tantalising 'what if' to ponder. Einstein's 1939 letter to President Roosevelt, pointing out that nuclear chain reactions could be used to devastating effect, is often credited with playing a key role in the US winning the race to build the atom bomb.

In truth, it did nothing of the sort. Roosevelt's administration did not start thinking seriously about the bomb until 1941, when a paper by two German physicists living in the UK, Rudolf Peierls and Otto Frisch, demonstrated that one might be possible. And in any case Nazi scientists under Werner Heisenberg never managed to make a bomb, nor even a working reactor. Einstein's non-intervention would have made no difference.

What if the Nazis had won World War II?

? If the result of the second world war had been different, the scientific agenda for the fifty years after would have been dominated not by subatomic physics and nuclear energy, but by a sinister form of environmentalism, says Steve Fuller.

In early 1941, the Nazis invaded Russia, a disastrous decision that ultimately cost them the second world war. But it wasn't the only course they could have taken. As historian John Keegan points out in a 1999 essay 'How Hitler could have won the war', the Nazis could easily have chosen to conquer the Middle East's oilfields instead. Even if this had not been entirely successful, Hitler would have probably ended up controlling enough of Europe's energy supplies to force a stalemate, ending the war two or three years early. This outcome would have prevented most – if not all – of the Holocaust, which may have been inspired by the cosmic approval that Hitler read into his early Russian victories. The consequences for science would also have been profound.

Had the Nazis won (or at least not lost), the scientific agenda of the following half-century would have been dominated not by subatomic physics and nuclear energy, but by ecology. Ideas such as biodiversity, the precautionary principle and animal rights would have been the dominant concepts of a political form of social Darwinism, built on the tenets of racial hygiene.

At first sight this seems an unpalatable conclusion. It is hard to believe that the success of Nazism could have given rise to a world with any redeeming features. But even in the real world, the Nazi defeat did not stop much of their science from being assimilated by the victor nations.

Suppose, then, a 1943 peace treaty allowed Hitler to retain his European and Asian conquests. Nazi economists, aware of Germany's lack of natural resources, would have demanded a re-agrarianisation of conquered nations to prevent them from becoming competitors. Command over at least some of the Middle East's oil would have allowed the Nazis to limit the pace of competition among the remaining free nations. The Nazi empire would thus have become a global superpower.

What would that have meant for science and technology? The ideology of racial hygiene – which pre-dated Hitler's rise and declined only with his fall – took Earth's point of view, nowadays popularised as Gaia, with deadly seriousness. Racial hygienists held, for example, that global misery resulted from misguided human attempts to reverse the effects of natural selection. Thus, one important result would have been the end of mass immunisation, which the Nazis considered emblematic of 'counter-selection'. For racial hygienists, vaccines did not restore the body to a natural state, but artificially enhanced the body. Vaccine research had also historically been driven by the mixing of peoples caused by imperial expansion, which led racial hygienists to conclude that only states with stable and 'pure' populations could survive naturally. The implications for medical research and policy would be clear. The Nazis would have omitted vaccines from what we now call preventive medicine, a field in which they were otherwise pioneers.

This interest in preventive medicine, however, meant that research into the health effects of radiation, asbestos, heavy metals, alcohol and tobacco would have advanced more rapidly. The Nazis would have also mandated the production of organic foods, outlawed vivisection and encouraged vegetarianism and natural healing. What is more, the eco-friendly Nazis' sensitivity to the scarcity of the world's oil supply would have sparked an early scientific interest in curtailing carbon emissions and shifting to alternative energy. In short, the late 1940s would have seen scientifically informed policies that only began to be pursued for real in the late 1960s.

There would also have been compulsory sterilisation and permissible euthanasia, done in the name of reversing the 'damage' caused to the human ecosystem by those nineteenth-century enemies of biodiversity, the bacteriologists Louis Pasteur and Robert Koch, who failed to grasp that, say, tuberculosis was nature's way of culling an unsustainable human population. Over time, as a balance to nature was deemed to have been restored, sterilisation and euthanasia might no longer have been required.

All of these developments presuppose a state-enforced 'corporate environmentalism' that would have reached an early accommodation between big business and the environment. In the process, however, the value of human life would have become negotiable. Those who raised objections to the natural selection of *Homo sapiens* would be consigned to the political and scientific margins. The centre ground would be occupied by debates over whether the culling of humans should be an active or passive process.

The Nazis would also have pioneered the first manned space missions. They would have realised that sending surplus people into space might enable them both to test the limits of their most advanced physical sciences – astrophysics and aeronautics – and to expand the Reich's carrying capacity to other planets or orbiting space stations. The latter would have come to be seen as a humane yet informed alternative to culling.

Finally, what of nuclear physics? An early end to the war would have halted the race to build the atomic bomb, which the Nazis had undertaken grudgingly in response to the Manhattan Project. And with much of the ecosystem under direct political control, there would be little need to research nuclear energy. The very idea of smashing atoms to release untold energy, as outlined in Albert Einstein's letter of 2 August 1939 encouraging President Roosevelt along these lines, would have been used to stoke the flames of anti-Semitism. Jews would have been demonised for having recommended a bomb that upon explosion would have brought about a different but equally lethal final solution.

What if we'd never stopped travelling to the moon?

? Life on Earth could have turned out differently if the Apollo missions had continued, but what about life off Earth, ponders Henry Spencer.

It is 14 December 1972. Eugene Cernan and Harrison Schmitt climb back aboard *Apollo 17*'s lunar module. Just three years after Neil Armstrong's historic step, human exploration of the moon is coming to an end.

The die had been cast years before *Apollo 11* had even reached the moon. In the late 1960s, the Vietnam war was straining US finances. A fatal fire on the Apollo launch pad in January 1967 had blotted NASA's copybook. The Soviet moon effort seemed to be going nowhere. In the budget debates during the summer of 1967, Congress refused NASA's request to fund an extended moon programme. What if things had been different that summer? Suppose Congress had granted NASA's wish, then fast-forward forty-odd years . . .

It's July 2018, and in Johnson City, America's permanent colony on the moon – named after Lyndon B. Johnson, the president who authorised it – they are celebrating the third generation of lunar Americans: the first child born to parents themselves born on the moon. With just 5,000 inhabitants, 'city' is perhaps too grandiose a term. Those who had anticipated a domed city of the kind that once graced science fiction comics had also been disappointed. That idea never stood up to the harsh lunar reality of cosmic rays and meteorite bombardment.

Fortunately, there was a ready-made alternative: subsurface lava tubes carved by extinct volcanic flows. The early Apollo missions had seen evidence for partly collapsed tubes in the form of intermittent surface trenches. The extreme dryness of the rock and the moon's weak gravity meant tubes a kilometre or more in diameter could remain stable, and it proved a straightforward task to seal off and pressurise one of these in the late 1970s.

In the beginning, the government had run almost everything in Johnson City. These days, though, lunar exploration increasingly depends on the colony's independent businesses. Shipping costs of material from Earth are still astronomical, and local extraction of water and oxygen is a boom business. There's talk of a trade in mined helium-3 for use in fusion reactors on Earth – if fusion technology ever gets established on the mother planet.

There are thriving schools for the colony's children, but more advanced education is beamed up from Earth. Forty years on from Apollo, passenger flights into space are routine, but travelling to Earth is still expensive, and lunar inhabitants can't easily adapt to its higher gravity. Even if, like 90 per cent of the population, you were born on Earth, you can't easily go back. These problems are not shared by space tourists, who pay a pretty penny for a week or so's stay in orbit – and soon in the first hotel in Johnson City.

There's a Russian lunar colony too, a legacy of Nikita Khrushchev's desire to establish a permanent moon presence for the Soviet Union. In fact, the moon in 2009 is more multinational than ever: European and Japanese voices fill Johnson City, and China and India are considering starting their own colonies.

Exploring beyond the moon has been less of a success. In 1980, three US astronauts landed on Mars for a stay of a week, but that was a hastily conceived 'flags and footprints' effort to keep ahead of the Soviet competition. A Soviet landing and a couple more US missions followed, but by the late 1980s crewed missions to Mars had stopped. NASA's fixed budget was strained by cost overruns in establishing Johnson City, and the tottering Soviet economy was unable to match even that.

So apart from a visit to two near-Earth asteroids using leftover Mars hardware in the early 1990s, exploration beyond the moon stopped until the turn of the millennium. Technological improvements then inspired a cautious revival. Not least of these was the ability to supply rocket fuel extracted from the moon's unexpectedly rich polar ice deposits, meaning long-distance missions could be launched far more cheaply from the moon than from Earth.

In January 2004, a team of four Americans and two Europeans made it back to Mars, spending fourteen months on the first extended surface exploration. They found spectacular scenery, and the odd hint of past riches: traces of organic compounds, tiny rock features that might be microfossils. But life on Mars was clearly never common. If it is still around, it probably only inhabits niches such as the neighbourhood of geothermal vents – if they exist.

Analysing geology from orbit is always tricky, and the one hoped-for geothermal area within reach of the 2004 Mars explorers turned out to be a damp squib. It didn't help that their deep-drilling rig didn't work well in the unexpectedly hard and abrasive Martian permafrost. Still, hopes are high that the next multinational mission, to be launched in 2011, will be followed by a new base and perhaps even a colony.

All in all, humanity's engagement with space has been a positive affair. Launching and repairing communications satellites has become routine. High-speed computer networking and applications such as remote learning and telemedicine – developed for the early Mars expeditions – have brought spin-off benefits back on Earth.

The scarce resources that characterise space exploration and colonisation have boosted green technologies such as waste water recycling, which is finding Earthly use as we try to cool our warming world. The same goes for a prototype orbiting solar power station currently under construction. Its use of a microwave beam to transmit power to Earth was controversial, but the worries have been trumped by the prospect of abundant clean energy.

Progress in space has been slower than in the dreams of the pioneers of the early 1960s. But with a permanent presence on the moon, and plans for bases farther afield, in 2018 the idea is firmly entrenched that, for humanity, the sky is no limit.

Plot a course for page 187 to find out why life on Mars will be no picnic for early colonists.

3 The Fork in the Road

Ruminations on the multiverse tend to focus on key events in the past that could have sent us spiralling into a different timeline. If only we'd known then what we do now, the lament goes, we might have chosen to head down a different path.

Fortunately, the multiverse isn't done branching yet. We can still exercise some control over which future universe we end up in. This chapter looks at the discoveries, actions and events that could offer us a new path to follow, ones that will have our descendants looking back and wondering how it might have turned out differently. By studying the possible outcomes of each, we can inform ourselves of what's at stake. And increase the odds that our descendants will feel we made the right choice.

What if we find ET?

? In 1959, physicists Giuseppe Cocconi and Philip Morrison suggested that radio telescopes had become sensitive enough to detect signals emanating from alien worlds. Frank Drake made the first attempt the following year, and we've been searching ever since. MacGregor Campbell channels a discovery that would strike the final blow to the idea that we're the centre of the cosmos.

Thanks to the Kepler Space Telescope, we know the galaxy could hold as many as 30 billion planets similar to our own. The next generation of eyes in the sky, such as the James Webb Space Telescope, slated to launch in 2018, will search the atmospheres of such exoplanets for signs of life. Some think it's just a matter of time before we find out we're not alone. In 2015, NASA's chief scientist, Ellen Stofan, predicted we would have 'strong indications of life' on other planets by 2025. If she is right, how will we deal with the news?

What we detect will make a big difference to how we react, says Steven J. Dick, a former NASA historian and current astrobiology chair at the US Library of Congress in Washington, DC. Any discovery that is less obvious than little green men landing during the World Cup final is likely to be met by years of questions and examination. Sara Seager, a planetary scientist at the Massachusetts Institute of Technology who is searching for another Earth, agrees. It will probably take time to confirm any initial findings, she says. 'There may not be an "aha" moment.'

A chemical imbalance in an exoplanet's atmosphere could be a sign of microbial activity. But an indirect result such as this will probably have only a short-term impact, says Dick. The apparent discovery of Martian nanofossils in meteorite ALH84001 in 1996 led to a media frenzy, and even US congressional hearings, before the furore died down in the face of increasing scepticism. Most now think that the meteorite does not hold the remnants of ancient alien life.

A decoded broadcast from intelligent aliens would be altogether different. Scientists and governments would have to assess whether the message was threatening and what, if anything, should be sent as a response. It would pose a challenge to certain religions too, says Dick. 'Does Jesus have to be a planet-hopping saviour to all ETs?' Some people might see intelligent aliens as saviours themselves, giving rise to new religions. Others may simply celebrate them as species that overcame their provincial squabbles to explore the universe.

In the longer term, even slight evidence of extraterrestrial life would spark a quest to understand the universal principles of biology, says Dick. We may find answers to questions such as: Does life arise wherever the conditions are right, or is it a freak accident? Are there other types of genetic code? Does life always require carbon or water? Is Darwinian natural selection a universal, or are there other forms of evolution?

Perhaps most significantly, it would be the decisive blow to the idea that humans are the centre of the cosmos or the reason for its existence. Instead, we would be forced to acknowledge our place as just one tiny branch of a vast galactic tree of life. 'I hope that people will find a new sense of peace and an understanding that we are not alone,' says Seager.

. . . And what if we don't?

? All our searches for extraterrestrial life have turned up nothing. If we really are alone in the universe, says Michael Brooks, should we take Earth's life to other planets? Especially as we might already have the means.

It's fun to speculate about aliens. But what if there are no aliens? It's been sixty-five years since Enrico Fermi first pointed out our solitude. Fermi estimated that it would take an advanced technological civilisation 10 million years or so to fill the galaxy with its spawn. Our galaxy is 10,000 times older than that. Where is everybody?

It's not as though we haven't been looking. Not for long, perhaps, and not very hard, but even a crude estimate suggests there should be other advanced civilisations capable of signalling over interstellar distances. And yet – nothing.

So what if we really are alone, or so isolated as to amount to the same thing? 'If we think we are the only life in the universe, we have a huge responsibility to spread life to the stars,' says Anders Sandberg of the University of Oxford's Future of Humanity Institute. 'If we are the only intelligence, we may have an almost equal responsibility to spread that, too.'

NASA astronomer David Grinspoon agrees, although he hasn't given up on finding ET yet. 'We have these powers that no other species has had before,' he says. 'If we are it, if we are the best the universe has got, if we are the universe's sole repository of intelligence and wisdom and scientific insight and technology, it ups the ante quite a bit. We have a responsibility to preserve our civilisation.'

It won't be easy. First, we need to decide where to boldly go. We don't know if humans can survive for any meaningful length of time anywhere except the surface of Earth. 'Nowhere in our solar system offers an environment even as clement as the Antarctic or the top of Everest,' says the UK's Astronomer Royal, Martin Rees. But some pioneers aim to give it a go anyway. Billionaire inventor Elon Musk is aiming to establish a self-sustaining colony on Mars in the next fifty years. 'By 2100, groups of pioneers may have established bases entirely independent from Earth,' Rees says.

Second, we are going to need some serious propulsion power but we don't yet know what that will look like. Third, we have to have some way to deal with interstellar dust, which could create catastrophic collisions with our craft at the speeds we would need to attain. Fourth, we would need some kind of artificial gravity onboard, otherwise the crew will suffer massive, possibly fatal, health issues.

There will undoubtedly be many more obstacles that we have yet to confront or even imagine. But Sandberg and others are optimistic we can overcome them.

Even if we can't, we could play the longer game and attempt to seed the galaxy with life via 'directed panspermia'. The basic idea is to launch microorganisms into space in the hope that they will crash-land on a planet or moon suitable for life, and eventually evolve into a self-aware, intelligent species.

Science fiction author Charlie Stross suggests isolating spore-forming archaea and photosynthetic bacteria that can survive for long periods in the harshest of environments. 'Put them on rockets and fire them out of the solar system,' he says. 'Almost all will perish, but if you launch a hundred tonnes of spores every year for a century, maybe sooner or later something will work.'

There would be no real payback for us, except perhaps returning a favour. Life on Earth may have started by directed panspermia. If so, our ultimate purpose may be to pass it on again, like a chain letter through the cosmos.

What if we could see the future?

? From Laplace's demon time to places where time flows uphill, the multiverse offers many chances to see what's coming. Gilead Amit asks what it would be like to have such clairvoyance, while Joshua Sokol explains why humans might be biologically destined to only travel through time looking backwards.

Human beings are adept mental time travellers. Our ability to envisage how things might be in the future, which seems unmatched by any other species, is arguably what has made us the cultured and civilised animals we are. But if visualising possible futures is a game-changer, being able to predict the future would be nothing short of revolutionary.

Debate has long raged about whether that is even possible. According to one school of scientific thought, known as determinism, it is. Given enough data about each atom in the universe, we can know tomorrow's football scores with as much certainty as yesterday's. This mindset suffered a couple of blows during the twentieth century. First, Heisenberg's notorious uncertainty principle said it was impossible to know everything about a quantum system such as an atom. Second, chaos theory taught us that the future behaviour of any physical system is extraordinarily sensitive to small changes – the flap of a butterfly's wings can set off a hurricane, as the saying goes.

But even if it's theoretically impossible, in practice we might get as close as makes no difference. Computers are already producing ever more accurate simulations of future reality, from tomorrow's weather to long-term climate trends to the eventual fate of our galaxy. Extrapolating from current number-crunching capabilities, near-perfect climate prediction, for example, should be possible within a century or so, says climate scientist Gavin Schmidt of NASA's Goddard Institute for Space Studies in New York City.

Such soothsaying ability might not play to our advantage, says Matteo Mameli, a philosopher at King's College London. Predictive software might ultimately deprive us of that evolutionarily hard-won ability to think creatively and improvise our way out of dangerous situations. Alternatively, unrestrained by any fear of failure, our hubris may accelerate our destruction of the world around us.

The outcome might depend on who has access to the

predictive tools, says Timothy Pleskac, a psychologist at the Max Planck Institute for Human Development in Berlin, Germany. In the wrong hands, they might help prop up dictatorships or establish commercial monopolies. But more socially minded governments could use them to ready their citizens for challenges such as approaching environmental disasters.

Or, Pleskac thinks, our supremely adaptive minds might finally find themselves overwhelmed by such omniscience – and reject it in favour of a quiet life. 'They might say, all that information is there, but I don't want to have access to it,' says Pleskac. 'It may be quite adaptive to be ignorant. People may just want to be left alone.'

Why the future keeps slipping your mind

Remember to keep your past and future in the proper order. Although the laws of physics don't care if you live backwards in time, natural selection might. Heinrich Päs of the Technical University of Dortmund, Germany, and his team modelled the lives of 'agents' trying to dodge falling rocks within a two-dimensional grid. 'We have something like sex and violence. The rocks kill the individuals, and the individuals breed,' Päs says.

Each rock bounces off the grid's walls like a ball on a billiard table. To give the simulation an arrow of time, rocks break in half on hitting a wall, which increases entropy, a measure of time's flow. If an agent is about to get hit by a rock, they have some chance of stepping out of the way, but they can only avoid one at a time. If two are incoming, they will certainly be hit, draining their 'health' by an amount determined by the size of the rock.

Päs's team ran the simulation with agents that look either

forwards in time or backwards, against entropy. They found that agent populations grew better when they saw time run forwards. That's because when time runs forwards, agents only have to predict the split of one rock to stay safe. When it runs backwards, they must track many little rocks coming together into one.

That's not too surprising, says James Hartle of the University of California in Santa Barbara. It has been shown that laying down memories involves an overall increase of entropy. By contrast, systems with unchanging entropy, like a frictionless swinging pendulum, can't act as memory stores. So of course memories can only come from the past, Hartle says.

Still, Päs thinks the potential evolutionary benefit of mentally tracking entropy's increase might help shape the way we perceive the past and future. 'It looks like a crazy idea to use some scenario of evolution to explain some very basic property of time itself,' he says. 'The point is more that you really can do it.'

. . . But could do nothing about it?

? Determinism might allow us to see the future, but it strikes a blow against free will. Our sense of justice assumes we are free to choose our actions. If that turns out to be an illusion, society could suffer, finds Sean O'Neill.

Our moral sense is based on an assumption so fundamental it seems unassailable: that we are masters of our own destiny. But the more we unpick the subtle knot tying conscious experience to the brain, the shakier that assumption feels.

The free will debate is an old one, but in the 1980s psychologist Benjamin Libet really stirred things up. He performed an experiment that revealed a signal in the brain moments before his subjects felt a conscious intention to move a finger. Although Libet's work remains controversial, it raises a big question: is our unconscious brain really in the driver's seat, with our consciousness a mere passenger?

Even if we're driving, we may be on rails. Split-second life-or-death decisions – a police officer choosing to fire a gun, say – are often made too quickly for conscious deliberation to play a part. Instead, such choices may be guided by hardwired, unconscious bias.

What if a breakthrough in neuroscience stripped us of free will? One outcome may be a loosening of morals. In experiments, people behave more selfishly and dishonestly if they are persuaded beforehand that free will is largely an illusion. They are also more likely to treat wrongdoers leniently, offering a hypothetical criminal a shorter prison sentence than they would otherwise have done. It's harder to ascribe blame to an automaton, after all. But these behavioural changes only last until the powerful feeling of our own agency reasserts itself.

On the other hand, belief in free will tends to be strengthened by considering a scenario in which someone acts immorally. Joshua Knobe of Yale University and his colleagues argue that our powerful belief in free will is bound up with a fundamental desire to hold others responsible for their harmful actions. In other words, belief in free will is required to justify punishment. And there is some evidence that fear of punishment is what keeps societies from breaking down.

Legal punishment for the purpose of retribution would make much less sense if the idea of free will were jettisoned, says Azim Shariff at the University of Oregon in Eugene.

But it might still serve a practical need as a deterrent. 'Just as we make efforts to avoid the negative consequences of other natural phenomena like hurricanes and rat infestations, so too can we make efforts to incapacitate transgressors to stop them causing further harm,' he says.

So we would probably keep our legal systems in place. Faced with the prospect of punishment, our brains work to keep us out of jail – consciously or unconsciously.

The loss of free will may be a harder pill to swallow when things get personal and our emotions come into play. 'Thinking about a guy being mean to your sister is likely to motivate people to reject that neuroscience breakthrough,' says Shariff.

Knobe agrees. 'If people discovered that there was no free will, they would undoubtedly come to think very differently at an abstract level,' he says. 'Yet research suggests this sort of abstract reflection would have shockingly little impact on the way people actually treated each other.'

In any case, hanging on to a strong belief in our own agency has its upsides. It is linked to a greater sense of satisfaction and self-efficacy, higher commitment in relationships and greater meaningfulness in life. Free will is dead? Long live free will!

What if we decide that God exists?

? As shocks to the system go, discovering evidence for God might be the biggest, says Joshua Howegogo. But it might also spell the end for religion as we know it.

The equations are checked and rechecked. Finally, physicists throw up their hands and declare that the big bang must have had a cause – a prime mover that created the universe. Or perhaps God simply shows up on Earth in full supernatural glory. As shocks to the system go, it couldn't get much bigger.

There's no denying we have a God-shaped void in our heads. What if it were filled? Would our increasingly secular world see a mass conversion? Perhaps, but it is far from clear what people would convert to. In fact, organised religions would probably be thrown into disarray. If God could come in any number of guises, whatever this entity turned out to be would be unlikely to fit our narrow range of existing ideas.

We might also think that proof of God would be egg on the face of atheists. Perhaps – but for many, the idea of God is not only unbelievable, but also distasteful. The writer Christopher Hitchens, who called himself an 'anti-theist', detested the idea of a cosmic chaperone watching our every move. If this being did present itself to us, we could see atheists start a revolution against God.

Ultimately, our reaction would depend on the nature of the evidence, says Joel Robbins, an anthropologist of religion at the University of Cambridge. Big changes would only come from a dramatic confirmation, he says. 'It would take something like Jesus or aliens turning up on Earth and saying, "Hi, I'm your creator".'

Although religion might take a hit, we could expect a boom in theological debate. At the top of the agenda would be morality, suffering and death – subjects often sidelined to academic backwaters, says Stephen Bullivant at St Mary's University in London. For example, if suffering exists despite there being an all-powerful God, does that mean suffering is somehow necessary in the universe? Or does it instead say something about the nature of this God?

There may also be a swing towards fatalism. There is evidence that religious people often simultaneously invoke natural and supernatural explanations for events such as sickness and death. People will take medicine even when they believe God is controlling the course of their illness, for example. Proof of God's existence might tip the balance away from medical explanations of sickness and encourage people to feel that when you get ill, that's it – your time has come.

Knowing that a physical entity set off the big bang would help cosmologists, says Alexei Nesteruk, a cosmologist and theologist at the University of Portsmouth, UK. It would prove once and for all that the universe is not eternal.

There are two broad theories that suggest the universe isn't eternal. Inflationary cosmology says that universes pop up like bubbles within a high-energy vacuum with repulsive gravity. And conformal cyclic cosmology says that parallel universes continually collide, move apart and collide again, each collision seeming like a big bang. A God might shed light on which of the competing models of our universe – or multiverse – is correct.

Ultimately, though, Nesteruk thinks proof of God is impossible, if we take this to mean evidence of an uncreated being. Faced with such an entity, we would always ask, 'Well, who created you?'

. . . And could we have religion without one?

? Few organisations have been as effective at bonding large groups of humans – or turning them against each other – as religions. Kate Douglas wonders if the solution is to offer all that's super about religion, without the supernatural.

What form would the ideal religion take? Some might argue that instead of redesigning religion, we should get rid of it. But it is good for some things: religious people are happier and healthier, and religion offers community. Besides, secularism has passed its zenith, according to Jon Lanman, who studies atheism at Queens University Belfast. In a globalised world, he says, migrations and economic instability breed fear, and when people's values feel under threat, religion thrives.

Today's religions come in four flavours, according to Harvey Whitehouse at the University of Oxford. First, the 'sacred party', such as incense burning, bell ringing and celestial choral music in Catholicism. Second, 'therapy': for example, the practices of healing and casting out devils among some evangelical Christians. Third, 'mystical quest', such as the Buddhist quest for nirvana. And finally, 'school': detailed study of the Koran in Islam or reading the Torah in Judaism.

While each appeals to a different sort of person, they all tap into basic human needs and desires, so a new world religion would have a harmonious blend of them all: the euphoria and sensual trappings of a sacred party, the sympathy and soothing balms of therapy, the mysteries and revelations of an eternal journey and the nurturing, didactic atmosphere of a school.

Numerous festivals, holidays and rituals would keep followers hooked. 'Rites of terror' such as body mutilation are out – although they bind people together very intensely, they are not usually compatible with world religions. Still, highly rousing, traumatic rituals might still feature as initiation ceremonies, because people tend to be more committed to a religion and tolerant of its failings after paying a high price for entry.

The everyday rituals will focus on rhythmic dancing and

chanting to stimulate the release of endorphins, which Robin Dunbar, also at Oxford, says are key to social cohesion. To keep people coming back, he also prescribes 'some myths that break the laws of physics, but not too much', and no extreme mysticism, as it tends to lead to schisms.

With many gods and great tolerance of idiosyncratic local practices, the new religion will be highly adaptable to the needs of different congregations without losing its unifying identity. The religion will also emphasise worldly affairs – it would promote the use of contraceptives and small families and be big on environmental issues, philanthropy, pacifism and cooperation.

Now, what shall we call it? Utopianity?

What if we're not the only intelligent species on Earth?

? If we could communicate with other animals it would force us to take a long hard look at our relationship with them and the environment, says Daniel Cossins.

In 2015, a New York court ruled that Hercules and Leo, two research chimps at Stony Brook University, had no right to legal personhood. But the fact that such a case made it through the courts at all shows our new willingness to consider the issue of personhood for other species. 'Efforts to extend legal rights to chimpanzees . . . are understandable; some day they may even succeed,' wrote judge Barbara Jaffe.

Steven Wise, a lawyer at the Florida-based Nonhuman Rights Project, which brought the lawsuit, argues that if chimps are declared legal persons, they should be granted rights to protect their fundamental interests. 'That would certainly include bodily liberty and likely bodily integrity as well,' he says. We could no longer keep chimps in captivity, never mind subject them to intrusive experimental procedures.

If chimps were given rights, we might expect other intelligent species, such as killer whales and elephants, to follow. But why stop there? Our ideas about the inner lives of other animals – their capacity for suffering, autonomy and self-awareness – are based largely on analogy with ourselves: how would we like it in their place?

But what if those animals could tell us? What if a dog or dairy cow could let us know how it felt about its lot in life? The idea may not be as far-fetched as it seems. There are many examples of communication between apes and their human keepers. Researchers are busy decoding dolphin. And cognitive scientists are beginning to study emotional states in animals. It may only be a matter of time before more meaningful communication between species is possible.

Would we still eat meat once that happens? If we could converse with pigs, say, how could we justify slaughtering them by the billion, however humanely? And where should the line be redrawn? Would we still eat fish? Many of us might shun meat and animal products entirely.

Widespread legal rights for animals would affect environmental efforts too. Conservationists would have to put down the gun, says biologist Marc Bekoff, formerly of the University of Colorado in Boulder. Right now, most people take a utilitarian view, considering it acceptable to kill members of one species to save another or to safeguard an ecosystem. 'But if we accept that these animals are sentient beings and ascribe greater value to each individual's life, you have to come up with alternative strategies,' says Bekoff, a leading voice in the compassionate conservation movement. He insists that a 'do no harm' approach is possible, although others argue it would make us too sentimental to do much good.

How would we weigh an animal's life against a human one? Research on animals leads to treatments that save human lives, making a blanket ban on animal testing unlikely. But asking scientists to limit the pain and suffering they inflict will no longer be enough, says Bekoff. Scientists would have to make the case that the benefits for humans outweigh the harm to the animal. At the very least, lots more species would get their day in court.

Could we save the world by giving up meat? Skip to page 133 to find out.

. . . But intelligence is a dead end?

? We're smug about our smarts, says Anil Ananthaswamy. Our intellect separates us from other animals and got us to our position of planetary dominance. But it could also be our downfall – and the reason we've yet to find intelligent alien life.

Our intelligence, the very trait we like to think makes us the pinnacle of evolution, could be our undoing. 'Human beings tend to think being clever is such a good thing, but it might be that from an evolutionary perspective, being stupid is much better,' says philosopher Thomas Metzinger of the University of Mainz, Germany.

Humans have evolved a unique form of intelligence, with cognitive complexity unseen in other species. This has been the secret behind our agricultural, scientific and technological progress. It has let us dominate a planet and understand vast amounts about the universe. But it has also brought us to the brink of catastrophe: climate change looms and a mass extinction is already under way, yet there is little sign of a concerted effort to change our ways.

Our troubles could be compounded by the fact that human genetic diversity is abysmally low. 'One small group of chimpanzees has more genetic diversity than the entire human species,' says Michael Graziano of Princeton University. It's not unthinkable that a global disaster could wipe us out.

For this, we have an awkward double-act to blame. Metzinger argues that we have reached this point because our intellectual prowess must still work alongside hardwired primitive traits. 'It is cognitive complexity, but without compassion and flexibility in our motivational structure,' says Metzinger.

In other words, we are still motivated by some rather basic instincts, such as greed and jealousy, and not by a desire for global solidarity, empathy or rationality. And it's unclear whether we will evolve the necessary social skills in time to thwart planetary disaster.

Another part of the problem is that our intelligence comes with so-called cognitive biases. For instance, psychologists have shown that humans pay less attention to future risk compared with present risk, something that makes us routinely take decisions that are good in the short term but disastrous in the long term. This may be behind our inability to fully fathom the risks of climate change, for example.

Humans also have what philosophers call existence bias, which influences our view of the value of life – it's better to exist than not. Ultimately, we tend to focus on the positives. But what if our intelligence were to develop in a way that meant we lost such biases?

In fact, superintelligent aliens might have already achieved that. With a balanced outlook no longer weighted to the short term and a clear-eyed view of suffering, such a life form could decide that life is just not worth it. 'They may have come to the conclusion that it's better to terminate their own existence,' says Metzinger. Could that explain why we haven't yet made contact with an alien intelligence? 'Possibly,' he says.

4 Life, But Not as We Know It

Do we live in the best of all possible worlds? In his struggle to explain how suffering and injustice could exist in a universe created by an all-powerful, all-benevolent god, German polymath Gottfried Leibniz concluded we did. According to Leibniz's thinking, this god had imagined all possible universes – the multiverse if you will – and willed the best of the bunch into existence.

A relief perhaps for those unable to tour the rest of the multiverse, but it does leave a lot of problems here on Earth for us to tackle. In this chapter, we look at some of the biggest changes we could make to improve our world. Will we find that the road to hell is paved with these good intentions? Or could the best of all possible worlds be made even better?

What if we could start over?

? The way we live is mostly down to accidents of history. The architects of the modern world (and the ancient one) were forced to build without a plan, learning as they went. So what if we thought it through properly? Bob Holmes takes a tour of a more rational world.

In just a few thousand years, we humans have created a remarkable civilisation: cities, transport networks, governments, vast economies full of specialised labour and a host of cultural trappings. It all just about works, but it's hardly a model of rational design – instead, people in each generation have done the best they could with what they inherited from their predecessors. As a result, we've ended up trapped in what, in retrospect, look like mistakes. What sensible engineer, for example, would build a sprawling, low-density megalopolis like Los Angeles on purpose?

Suppose we could try again. Imagine that Civilisation 1.0 evaporated tomorrow, leaving us with unlimited manpower, a willing populace and – most important – all the knowledge we've accumulated about what works, what doesn't, and how we might avoid the errors we got locked into last time. If you had the chance to build Civilisation 2.0 from scratch, what would you do differently?

Redesigning civilisation is a tall order, and a complete blueprint would require many volumes, not just a few magazine pages – even if everybody agreed on everything. But, undaunted, *New Scientist* set out to discover what might be on the table, by seeking provocative ideas that challenge what we take for granted. The result is a recipe for overhauling how we live, get around, and organise our societies – as well as reconsidering our approach to concepts such as religion, democracy and even time. Dreaming of a new civilisation is more than a thought experiment: the answers highlight what is most in need of a rethink, and hint at bold repairs that might be possible today.

Take cities, for starters. Historically, they have generally arisen near resources that were important at the time – say harbours, farmland or minerals – and then grown higgledy-piggledy. Thus San Francisco developed on a superb harbour and got a boost from a mid-nineteenth century gold rush, while Paris grew from an easily defended island on a major river. How would we design cities without the constraints of historical development?

In many ways, the bigger cities are, the better. City dwellers have, on average, a smaller environmental footprint than those who live in smaller towns or rural areas. Indeed, when Geoffrey West of the Santa Fe Institute in New Mexico and his colleagues compared cities of different sizes, they found that doubling the size of a city leads to a 15 per cent decrease in the energy use per capita, the amount of roadway per capita, and other measures of resource use. For each doubling in size, city dwellers also benefit from a rise of around 15 per cent in income, wealth, the number of colleges, and other measures of socioeconomic well-being. Put simply, bigger cities do more with less.

Of course, there are limits to a city's size. For one thing, West notes, his study leaves out a crucial part of the equation: happiness. As cities grow, the increasing buzz that leads to greater productivity also quickens the pace of life. Crime, disease, even the average walking speed, also increase by 15 per cent per doubling of city size. 'That's not good, I suspect, for the individual,' he says. 'Keeping up on that treadmill, going faster and faster, may not reflect a better quality of life.'

But there's an even more fundamental limit to how big a city can get: no matter how efficiently its inhabitants use resources, a city must have a way to get enough food, materials and fresh water to support its population. 'Water is the most problematic of diminishing resources,' says Christopher Flavin, emeritus president at the Worldwatch Institute in Washington, DC. 'Oil can be replaced with renewable sources of energy. There are no good replacements for fresh water.'

No matter what the benefits of aggregation, then, our new civilisation is likely to need many cities of diverse sizes, each matched to the ability of the local environment to supply its needs. That means no megacities in the middle of the desert, like Phoenix, Arizona. Our larger cities should be close to good water sources, preferably along coasts to give access to energy-efficient shipping, and near fertile farmland. New York, Shanghai and Copenhagen all fit that bill; Los Angeles, Delhi and Beijing fall short.

Perhaps the biggest flaw of many cities is the suburb – the land-gobbling sprawl that creates communities far from shopping or commercial districts and forces people into their cars to travel. 'Urban sprawl has been a huge mistake,' says Flavin. It's been the dominant growth pattern of most North American cities, and is a major reason why Americans use so much more energy than Europeans, whose cities tend to mix residential and commercial uses in more walkable neighbourhoods.

Big cities like London and New York have already solved the car problem by making driving so impractical that most residents use mass transit, or walk or cycle. But even smaller cities could achieve this with the right design.

Mark Delucchi at the Institute of Transportation Studies at the University of California, Davis, envisions districts laid out concentrically around a central business hub which residents access on foot, by bicycle or with light vehicles like golf carts. 'We believe that one of the major things that keeps people out of these low-speed vehicles is that people don't feel they function safely enough in a regular road system,' he says. To avoid that, conventional cars and trucks would be segregated on separate roadways, perhaps at the outskirts of each district.

To make this layout practical, every resident would need to live within about three kilometres of a hub, Delucchi estimates, giving each district a population of about 50,000 to 100,000, while maintaining a pleasant living environment of low-rise buildings. Each hub could then link to other hubs through a mass transit system, allowing people easy access to other districts for work, and to the attractions of a larger city. A few cities, such as Milton Keynes in the UK and Masdar City in Abu Dhabi, already use some of these principles.

Once this basic structure was established on the large scale, much of the responsibility for design within each district could then be handed over to residents and local businesses. In a way, that's how cities used to evolve. For example, mills were set up by the river to take advantage of water power, then workers' houses were built within walking distance, while the mill owners built on the hills where the view was best. But over the past couple of centuries, this organic evolution has been replaced by top-down planning, leading to the sterile monotony of cities such as Brasilia in Brazil, and modern tract-housing suburbs.

Today, though, online social networking gives individual users tools to coordinate and cooperate like never before. 'I would build the cities in an open-source way, where everybody can actually participate to decide how it's used and how it changes,' says Carlo Ratti, an urban designer at the Massachusetts Institute of Technology. 'It's a similar process to what happens in Wikipedia.' By tapping into this sort of crowd-sourcing, the residents themselves could help plan their own wiki-neighbourhood, Ratti proposes. An entrepreneur seeking to start a sandwich shop, for example, could consult residents to find out where it is most needed. Likewise, developers and residents could collaborate in deciding the size, placement and amenities for a new housing block – even, perhaps, the placement of roads and walking paths.

With cities and transportation refashioned, the next problem our rebuilding society faces is energy. This one's easy: virtually everyone agrees the answer should be renewables. 'We can't say it should all be solar or it should all be wind. It's really critical that we have all of them,' says Lena Hansen, an electric system analyst with the Rocky Mountain Institute, an energy-efficiency think tank in Boulder, Colorado. That would help ensure a dependable supply. And instead of massive power plants, the best route would be small dispersed systems like rooftop solar panels. This decentralised generation system would be less vulnerable to extreme events like storms or attacks.

Hansen estimates that building an electricity system fully based on renewables, at least with our present technology, might cost a bit more upfront than recreating the present, fossil-fuel-based system, but fuel savings would quickly recoup that. Still, it might not be such a bad thing if energy was more expensive in our new civilisation, says Joseph Tainter, a sustainability scientist at Utah State University in Logan. Since energy is a cost in most manufacturing, cheap energy makes other material goods cheaper, too. 'It induces us to consume more and more – to produce more children, to consume other kinds of resources and let the society become more complex,' he says. To keep that from happening, Tainter suggests that energy prices might be kept artificially high.

An alternative might be to ensure prices for all goods reflect their true environmental costs. If the price of fossil fuels reflected the actual cost of global warming, for example, simple economics would push everyone towards radical improvements in energy efficiency and alternative energy sources.

While we're tinkering with the economy, we might want

to move away from using GDP as a measure of success. When nations began focusing on GDP after the second world war, it made sense to gauge an economy by its production of goods and services. 'At that time, what most people needed was stuff. They needed more food, better building structures – stuff that was lacking – to make them happy,' says Ida Kubiszewski of the Institute for Sustainable Solutions at Portland State University in Oregon. 'Now times have changed. That's no longer the limiting factor to happiness.'

Instead, we may want to broaden our indicator to include environmental quality, leisure time and human happiness – a trend a few governments are already considering. With Gross Domestic Happiness as our guide, people might be more likely to use gains in productivity to reduce their work hours rather than increase their salaries. That may sound utopian, but at least some societies routinely put greater value on happiness than on material things – such as the kingdom of Bhutan and the indigenous potlatch cultures of the west coast of North America that redistribute their property. 'I don't think it's contrary to human nature to have a system like this,' says Robert Costanza, an ecological economist also at Portland State.

After the economy, the next issue that needs to be dealt with in the new civilisation is the matter of government. We'll assume that some form of democracy is best, though there might be some discussion about the details. But the bigger question is, how many separate states would we want? Here, not surprisingly, opinions differ widely.

On one hand, humans evolved in small bands, and we still respond to challenges best in relatively small groups such as units of about 150, notes Robin Dunbar, an anthropologist at the University of Oxford. Governmental units no larger than, say, a Swiss canton would maintain this sense of commitment and local control in a way that is lost in larger units, he says.

On the other hand, increases in mobility, communication and technology – as well as the sheer size of the human population – mean that many of the world's problems are now truly global. 'What if there were a newspaper that was published just once a decade? What is the macroheadline of our time?' asks Paul Raskin, president of the Tellus Institute, a think tank in Boston. 'This decadal *New York Times* would be tracking a really major story, and it would have a headline something like "History has entered the planetary phase".' Just as events drove medieval city-states to amalgamate into nations centuries ago, global problems are now pressing for global solutions, he says. And that requires some form of global governance, at least to set broad goals – biodiversity standards, say, or global emissions caps – towards which local governments can find their own solutions.

All our design efforts to this point have been aimed at creating a sustainable, equitable and workable new civilisation. But if we want our new society to last through the ages, many sustainability researchers stress one more point: be careful not to make it too efficient.

The history of civilisations such as the Roman Empire or the Mayans suggests that they expanded dramatically during periods of climatic stability. Rulers knew how much they could get away with – how many fields they could irrigate from a single canal, for example, or how much forest to leave for the next generation of builders. That worked, and the civilisation flourished, until climate shifted. 'They ended up building themselves to a point that might have been very efficient, but when the environment started working differently, they had overbuilt,' says Scott Heckbert, an environmental economist at Alberta Innovates in Edmonton,

Canada, who simulates the collapse of past empires and peoples.

In the end, though, no human civilisation can last for ever. Every society encounters problems and solves them in whatever way seems most expedient, and every time it does so, it ratchets up its complexity – and its vulnerability. 'You can never fully anticipate the consequences of what you do,' notes Tainter. Every civilisation sows the seeds of its own eventual doom – and no matter how carefully we plan our new built-from-scratch civilisation, the most we can hope for is to delay the inevitable.

Could 100 babies left on an island rebuild civilisation?

? Without language, culture or tools, what would human children become, and how would their own children evolve? Christopher Kemp embarks on a forbidden experiment.

The island is a strange place. Overgrown, unpredictable, war-torn. For hours during the day, the sun climbs into the sky and it is quiet and peaceful. But later, as the shadows gather in the trees, a volley of hoots erupts from the forest canopy and echoes around the island. Moments later, an answering call bursts from a thickly wooded valley on the other side of the island. A call, and then a response. And then silence.

The sound comes again across the tops of the trees. Hooting, and then distant replies. High-pitched and repetitive, the sounds are not words. But they mean something anyway: the hunters are coming home. They emerge one by one from the foliage, stepping out cautiously into a wide and sandy bay. There are five of them, all males. Their bodies are lean and powerful. They carry a few simple tools – hammerstones and sharpened sticks and bones.

At the end of the bay, they meet another group of hunters. Gesturing to one another and making primitive sounds, the two groups become one. They make their way to an encampment near the treeline where the women and children wait for them. This is one of the tribes on the island. But the island is not real, and neither are the hunters. There is no encampment or waiting group of women and children. They are all just outcomes of a thought experiment.

It goes like this: many years ago, a cold-hearted and amoral scientist placed 100 babies on an uninhabited but fertile island, half of them boys, half girls. He provided only the minimum requirements to keep them alive. He left them food and water, being careful not to be seen. He kept them from harm, when possible. For years, the children received none of the trappings of a normal upbringing: no language, no education, no culture. Later, he slowly began feeding and watering them less and less, until eventually he gave them nothing at all. After twenty years, what have they become? How different are they from us? Are they merely hairless apes, or have they retained qualities that make them unquestionably human? Or, to put it more succinctly, when humans grow up without culture, do they invent it anyway?

In the six-million-plus years since the human lineage split from chimpanzees, evolution has endowed us with many of the attributes that make us who we are: bipedalism, hairlessness, opposable thumbs, extended childhood and a large and complex brain. But these features alone do not make us human. Many of our defining traits – such as language, art, technology, storytelling and cooking – are transmitted culturally. Although products of our biology, they are not fully encoded by genes. Instead, they pass from generation to generation by social learning, evolving as they go.

The relative contributions of these two forces – biological and cultural evolution, also known as 'nature' and 'nurture' – have been debated for centuries. How much of our humanity is hardwired, and how much of it depends on the culture in which we are raised? Are language and religion innate, for example? Are we born violent?

Disentangling biology and culture is difficult. They interact and reinforce one another. But there is an experiment that would help tease them apart. It can't be done in the real world, for ethical reasons. But it is possible to speculate, in the form of a thought experiment. Welcome to the island.

So after twenty years, what have those babies become? We cannot know for sure. But our thought experiment can draw on work from various scientific disciplines, including studies of hunter-gatherers, the evolution of birdsong, the development of sign language and research on children raised in orphanages. No one person can supply a complete answer, and many scientists are unwilling even to speculate. Some called the project wild and provocative, others fanciful and incoherent. One potential source dismissed the exercise as 'undergrad stoned talk'.

Others, though, were enthusiastic about imagining possible outcomes. 'My work is in language, so I've thought about this a lot,' says Ann Senghas, a psychologist at Barnard College in New York City who studies sign language in a population of deaf Nicaraguan children. The language emerged spontaneously in a school for the deaf that opened in Managua in 1977, and is now passed down from generation to generation, rapidly evolving like any spoken language.

It is linguistic gold dust because the signers have never learned a spoken language or to lip-read. But they invented

a language anyway, one that now has the complex linguistic hallmarks common to all others. This, says Senghas, supports the theory that the brain possesses structures that hardwire it for language. In the 1960s, linguist Noam Chomsky proposed that humans are born with a 'language acquisition device' – a hypothetical brain module that predisposes us to learn the languages in which we are immersed as babies and toddlers. And it seems that when the language acquisition device never encounters language, it invents one from scratch.

So would the first generation of island dwellers speak? 'Some people would think because so much of language is innate, that the first cohort would just have it right away, somehow,' Senghas says. 'But I don't think they would.'

Although they would, like the Nicaraguan children, invent ways to communicate. 'The first generation of those kids would develop certain ways of getting messages across non-verbally,' says Senghas. 'I'm betting they would at least come up with the idea of using gesture. I think they would probably use calls. There are obvious reasons that verbalisation is effective: you can call people from afar; you can alert many people at once.'

So, the island's first inhabitants, when they grow up, might well hoot to one another. Their vocalisations could travel across the treetops. Even more interesting, says Senghas, is the linguistic development she expects will take place in subsequent generations.

It will happen relatively quickly. Within a few generations, she says, the islanders will be speaking their own unique language. To support her claims, Senghas points to research on song traits in zebra finches, which suggests that the birds' vocalisation – and probably that of humans too – is genetic in origin, and later shaped by environmental factors. 'Only the male sings,' Senghas says, 'and the juvenile males learn their song from their fathers.'

If juvenile male zebra finches are raised in the absence of adult males, they fail to learn their signature song, but sing nonetheless. 'They develop some wacky, spooky song that doesn't sound like a natural zebra finch song,' says Senghas. 'It sounds a lot like white noise.'

When those juveniles reach adulthood and mate, they attempt to teach their offspring their strange discordant song. But something incredible happens: 'The next generation produces something a little more like natural zebra finch song,' says Senghas. And the generation after that produces something closer again. 'It only takes five generations before you've got the full-fledged thing.'

The first generation of island dwellers might not develop a language, but their brains have all the necessary structures and neural pathways. Like a juvenile zebra finch building on its father's atonal song, each successive generation of islanders will possess and develop more language than the last. 'I think it would take a couple of generations,' says Senghas, 'but not many to develop a language that's as rich and developed as anything that we have.' This is a crucial breakthrough for the islanders. Once in possession of language, cultural transmission of ideas becomes easier.

After they start to speak, says Ian Tattersall, a palaeoanthropologist at the American Museum of Natural History, they begin naming things. Eventually, the islanders begin to do something all human cultures do, and which probably emerged not long after sophisticated language did, perhaps 200,000 years ago: they name each other.

In many ways, our first generation is still resolutely human too. Even in the absence of culture, millions of years of biological evolution have endowed them with a complex brain and many distinctly human traits. They do not become mere apes overnight.

In fact, says Dominic Johnson at the University of Oxford, who uses evolutionary theory to study conflict and co-operation, if a Martian zoologist visited Earth to study its inhabitants, the islanders would provide better and more instructive examples of *Homo sapiens* than we do.

'Our physiology and behaviour were designed for life in the wilderness with scarce resources and a few dozen or so other people,' he says. 'That's our natural habitat. We are certainly not designed for life in modern megacities with fast food on every corner.'

In time, says Johnson, the islanders might begin to resemble modern hunter-gatherer groups. 'Hunter-gatherers provide a window into our human nature and the things that emerge spontaneously in human organisations when they're not surrounded by modern culture,' he says. For instance, the islanders are likely to invent tools and use them for many tasks. In the beginning, the tools are simple, developed by trial and error. Swung with determination, a heavy rock is a tool. A broken rock, with a sharp unfinished edge for cutting, is a better tool.

The island ecology will play its part too, says Johnson. Tribal groups in New Guinea and Tibet are vastly different to one another, their lives shaped fundamentally by their surroundings. For tools, the children can use only what they find on the island. Perhaps they will learn to fashion fish hooks from bone. They can harvest and eat only what grows there, and hunt only the animals endemic to the island and its waters. In time, by trial and error – and by making costly mistakes – they learn what is good and safe to eat, and they pass that knowledge on to each other and to their offspring.

They will be willing experimenters, and even more willing

learners. Humans seem to have an evolved tendency for individual learning and copying, especially of successful individuals. In this way, culture gradually improves and spreads, with new discoveries building on previous ones. Unlike groups that live in cold or mountainous places, the islanders go naked: they have no need to invent clothes.

Whatever their environment, the islanders will adapt to it, says Val Curtis, a behavioural scientist at the London School of Hygiene and Tropical Medicine and co-author of *Gaining Control: How Human Behavior Evolved*. 'Brains are learning machines,' says Curtis. 'Our minds come pre-equipped with structure that allows us to learn to behave adaptively.'

For example, if the children have to crack a nut, they intuitively understand that a rock will allow them to do it quickly. 'I predict that our abandoned kids will soon have their environment mapped and will figure out where to find or make shelters, beds and tools,' says Curtis. Each tool they develop will be a little more sophisticated than the previous one.

'But you know what?' says Senghas. 'It'll be a long time before they figure out the wheel.' The mastery of fire and the development of simple machines like ramps and pulleys will take thousands of years, says Senghas – the same amount of time as it took humans to invent them the first time. 'Generating things like levers and wheels and fire and cooking involved work and experimentation, and getting so far and then being able to pass that down culturally to the next generation,' she says.

There are other human traits that the children – and the adults they become – will pick up sooner. They laugh, says Curtis. They cry, sing and dance. They count, says Johnson – at least to two and perhaps higher. They don't recognise the concept of zero, which is a long leap into darkness.

Although they might appear feral in some ways to Westerners, they show disgust too, says Joshua Tybur, a social psychologist at VU University in Amsterdam, the Netherlands.

Disgust is an important protective mechanism, shielding us from potential dangers in our environment, says Tybur. 'Like many other species, we have adaptations for detecting pathogens and motivating pathogen avoidance,' he says. 'Disgust is one of those adaptations. I would bet everything I own on these children growing into adults who experience disgust.' These basic emotions and others – anger, joy, sadness, surprise and fear – are hardwired biological features that will emerge whether culture is present or not.

A never-ending battle is being waged between the tribes. Skirmishes. Ambushes. Raids. Attacks followed by counter-attacks. Battles sometimes last for days. Almost as soon as the babies were placed on the island, they began to form groups, says University of Chicago cultural anthropologist Richard Shweder. 'At least two groups and maybe more,' he says. Tribalism is something else that seems to be part of our biological make-up: it can be artificially induced in the lab by something as trivial as giving people different T-shirts, or separating people by eye colour.

The formation of groups facilitates cooperation, but also fuels conflict. 'People are going to trust members of their ingroup more than they're going to trust members of outgroups,' says Shweder. Groups won't share resources. If, for any reason, food or other materials on the island are limited, the tribes will clash over them.

Using traditional hunter-gatherers as a template, Johnson predicts the formation of several different tribes. The 100 babies represent a large group, he says, perhaps too large to stay cohesive for very long.

'Small-scale hunter-gatherer bands vary in size but are

typically a couple of dozen more or less related kin,' he says. 'This means our island is likely to end up in four or five groups, all in potential competition with each other.' The tribes will fight each other for everything: space, food, tools, dominance.

Even within a group, there will be hierarchies and divisions. 'They're going to start stigmatising people who don't follow the norms,' says Shweder. 'They're going to have banishment.' By banishing members who don't conform, says Shweder, the islanders begin to build a culture of cooperation, predictability and egalitarianism.

Groups will probably also fracture along gender lines, says Shweder. 'Males are going to have a very fleeting relationship to the reproductive act,' he says. 'It's going to be: have sex, ejaculate. The females are going to be left with the pregnancy.'

Males will probably steal women from each other, and from other groups. 'We can expect considerable competition among males for females,' says Johnson. Patriarchy, with males holding primary power and females being exchanged between groups, is a widespread feature of traditional human societies, says Shweder, especially sedentary ones.

It is likely that the islanders will have sex in private. Early on, says Paul Pettitt, an archaeologist at Durham University, UK, the island will be a sexually competitive environment with public displays of sexuality not dissimilar to those seen in chimpanzee troops. But this will change. 'As social complexity increases and individuals realise that pair bonding is advantageous – one gender gives birth to children and nurtures them and the other hunts and provides high-quality food – privacy would rapidly become a desired commodity,' says Pettitt.

When the women do give birth, their instincts will guide them and the babies will thrive, says Curtis. 'Their surprised mothers will figure out what to do to keep them from screaming, because a screaming baby is unrewarding and a happy one is good to be around.' The hard work of looking after vulnerable offspring will probably also produce a sexual division of labour, says Shweder, with males hunting and females taking on lower-risk tasks near their children, such as gathering vegetation.

The original population will grow quickly. The tribes will disperse to occupy different parts of the island. The fighting will commence. During periods of hardship, such as drought or flood, says Johnson – again using hunter-gatherers as a model – some of the tribes might come together, forming one larger group, and then splinter again when the hardship passes.

But they are never likely to exceed groups of about 150. That is the cognitive limit – known as the Dunbar number – on the number of people with whom any individual can maintain a stable social relationship. Eventually, each tribe on the island will develop its own language. And, between skirmishes and conflicts, the groups will begin to codify their own traditions. From nothing, culture will emerge.

The tribes will even begin to form their own religious beliefs, says psychologist Konika Banerjee of Yale University, who wrote a 2013 scientific paper called 'Would Tarzan believe in God? Conditions for the emergence of religious belief'. 'Would these babies grow up to have religion?' asks Banerjee. 'My guess is that, assuming some degree of typical social cognitive development, we might expect to see certain quasi-supernatural intuitions.'

As a part of early development, she says, cognitive biases begin to emerge in the way that children understand the world. 'Humans are a prolific creative species,' says Banerjee. 'We come up with all sorts of theories and explanations and ideas for making sense of the world.'

They may believe that entities such as rocks and trees were designed for a purpose, and that life events have a deeper meaning. In time, the islanders will begin to worship a god, perhaps many. It wouldn't happen quickly. But it would happen.

'Because of our theory of mind, we tend to assume that things happen for a reason,' says Johnson. 'But the big god idea – that would be hundreds of years down the road.'

Eventually, death comes to the island. 'Given that the kids would have brains like ours, it would not take long to realise their own mortality,' says Pettitt, who wrote a book called *The Paleolithic Origins of Human Burial*. 'Their imaginations would use religion to overcome the death anxiety and give them reassuring beliefs about the world to come.'

Based on what we know about our ancestors' response to death, it is likely that the islanders will quickly develop funerary rituals, says Pettitt. Deaths provoke a flurry of social activity. As the size of the group grows, the rituals become more elaborate and codified. Each tribe will have its own burial site. They are quiet, unvisited parts of the island. These are the deadlands.

'These could be places that are particularly dangerous, where carnivores are numerous, for example,' says Pettitt. 'Places where group memory preserves traces of deaths, or simply places that seem a bit spooky and on which the childhood imagination has got to work.'

In some cases, the islanders have covered the trees and rocks around the deadlands with colourful handprints, and festooned the tree limbs with jewellery made from abundant objects on the islands – polished shells, a dried and stiffened seahorse. Symbols cover everything. The islanders daub their own bodies with the same symbols as those covering the deadlands.

Within a few hundred years, the islanders have traditions and cultures. They look like us. Or an earlier version of us, at least.

This is only one of the possible outcomes. Among the multitudinous alternatives, there is another worth mentioning. It is absolute. Two decades later, the only proof that the experiment even took place is the occasional appearance of a small bone on the deserted coastline. Nothing else remains. The forests have fallen silent. The islanders are all dead.

Tattersall believes this is the likeliest outcome. 'None of them would have survived,' he says. 'You cannot think of human beings as independent of their culture and their society. This goes back a long, long way before we were human. It goes back millions and millions of years, back into our primate past, back into our primitive mammal past.' Even the most basic aspects of our cognitive development depend on being raised by linguistic, articulate parents, embedded within a rich and historical culture, says Tattersall.

Harvard University cognitive scientist Steven Pinker agrees. 'It's a pathological thought experiment,' he says. Humans need social contact: experiments on chimpanzees by US psychologist Harry Harlow in the 1960s showed that primates, when subjected to total social isolation, exhibit permanent disturbances in simple social behaviour.

Research on institutionalised orphans has shown that social interaction with adults and caregivers is a necessary component of normal development. Proximity just to one another will not save the islanders, says Pinker. Abandoned children who endure social deprivation perform worse in IQ tests, and score much lower in tests of cognitive, motor and behavioural development. In other words, even the innate abilities that children possess require a spectrum of different inputs in order to function properly.

The same seems to be true for 'feral' children who spend all or some of their childhood alone in the wild, but the literature is less clear, with most accounts heavily contested. 'It's like studying how fish swim by taking them out of water and watching them flop on the ground,' says Pinker.

There is no way to know the outcome. If Tattersall is right, the island is quiet and pristine once more. But perhaps, in the evening, as the sun dips towards the water, a hoot rises from deep in the forest, rolling above the treetops and then across the island, telling everyone that the hunters are coming home.

What if we could redesign the planet?

? We've built canals between oceans and tunnels under the sea. But some engineers are thinking bigger, reveals Michael Marshall. Much, much bigger.

They said it would never happen. Yet by the time you read this, work should have begun on a massive new canal to link the Atlantic and Pacific oceans. Building the 278-kilometre-long canal through Nicaragua will require moving billions of tonnes of earth and cost at least $50 billion. If it is eventually completed, it will be wider, deeper and three times as long as the Panama Canal. Its backers claim it will be the biggest engineering project in history. But it is certainly not the biggest ever suggested. 'All of us live in places that are engineered and designed,' says mega-engineering expert Stanley Brunn of the University of Kentucky in Lexington. So it's natural to dream even bigger, he says.

That may be true. But some of the schemes sound like the plans of Bond villains, such as flooding California's Death

Valley or nuking the isthmus of Panama. Others, like damming entire seas to generate hydroelectricity, are on a mind-boggling scale. Here are seven of the world's biggest schemes. Could we really go ahead with any of them? And should we?

Damming the Atlantic

It doesn't get much bigger than this. We could build a barrier across the Strait of Gibraltar, effectively turning the Atlantic into a huge dam reservoir. This was first proposed in the 1920s by German architect Herman Sörgel. With the flow of water into the Mediterranean reduced, the sea would begin to evaporate. Allowing it to fall by 200 metres would create 600,000 square kilometres of new land.

The environmental impacts of Atlantropa, as this plan is known, would of course be gargantuan. Perhaps most, er, damning of all, lowering the Med by 200 metres would raise sea level in the rest of the world by 1.35 metres. 'It's impossible in terms of the politics,' says Richard Cathcart, a real-estate adviser in Burbank, California, and a mega-projects enthusiast who has written several articles and books. 'Academics are actually afraid to talk about big ideas,' Cathcart says.

With sea level set to rise tens of metres over the coming centuries because of global warming, Cathcart thinks the idea of a dam across the Strait of Gibraltar is worth revisiting. Instead of lowering the Med, a dam could maintain it at its current level, saving low-lying farmland from the sea as well as cities such as Venice and Alexandria. Egypt in particular would benefit. As things stand, rising waters will swamp large parts of the Nile delta and displace millions of people by 2100.

Trans-Atlantic Aqueduct

Northern Africa could do with some more fresh water. The nearest potential source is the world's second-largest river, the Congo, but it flows through a volatile, dangerous region. So why not tap the world's largest river, the Amazon, instead? All you'd need is a pipe. A very long pipe.

The idea of piping water all the way across the Atlantic has been around since at least 1993, when Heinrich Hemmer put it forward in a journal devoted to flights of fancy. He envisaged a pipe 4,300 kilometres long, carrying 10,000 cubic metres of water per second, enough to irrigate 315,000 square kilometres.

There the matter rested until 2010, when Viorel Badescu, a physicist at the Polytechnic University of Bucharest in Romania, revisited the idea with Cathcart. They proposed to submerge a pipeline 100 metres below the surface, and anchor it to the seabed at regular intervals. The pipe would have to be at least 30 metres wide, and have up to twenty pumping stations to keep the water flowing. It would start offshore in the plume of fresh water from the Amazon – 'water that has been discarded by the continent of South America', as Cathcart puts it. All in all, he estimates that the pipeline would cost about $20 trillion. Residents of the Sahara, start saving now.

It might be wise to start a bit smaller – perhaps by piping fresh water 2,000 kilometres from lush Papua New Guinea to Queensland in Australia. In 2010, businessman Fred Ariel announced plans for a feasibility study into a $30 billion pipeline. In 2014, the PNG government approved the idea in principle, but Queensland has said the plan is not under 'active consideration'.

Flood the depressions

In 1905, irrigation engineers in California accidentally flooded a depression that lay below sea level. The result was the Salton Sea, the largest lake in the state. There have been many proposals over the decades for flooding other low-lying areas.

The prime candidate is the Qattara depression in north-west Egypt, which lies as deep as 130 metres below sea level. It consists of 19,000 square kilometres of sand dunes, salt marshes and salt pans. The idea is to flood it with seawater from the Mediterranean, just 50 kilometres to the north. Generating electricity is the main motive: if water flows in at the same rate as it evaporates, generation could continue indefinitely. The 'Qattara Sea' would become ever more saline, but surrounding areas might benefit from cooler, wetter weather.

The idea has been around since at least 1912, and the Egyptian government looked into it in the 1960s and 1970s. Few people live in the Qattara, so politically it is doable. The biggest problem is the sheer scale of the construction, which would require tunnels to go under a range of hills between the Mediterranean and the depression. One construction plan involved nuclear bombs. You may not be surprised that Egypt abandoned the idea.

Interest in the idea has revived recently thanks to Desertec – a plan to build a vast solar power plant in North Africa. Magdi Ragheb, a nuclear engineer at the University of Illinois at Urbana-Champaign, has proposed storing energy from Desertec by pumping seawater through a pipeline to storage facilities on top of the hills. When more electricity is needed, this water would be allowed to run down into the depression, turning turbines as it went. There would be no need for tunnels.

Flooding areas like California's Death Valley would also help offset sea level rise caused by climate change. But it is not worth doing for this reason alone: even if we flooded all of the world's major depressions, it would barely make a difference.

The Salton Sea, meanwhile, is not a great advert. It did thrive for decades, but it is now drying out and dying. Most fish can no longer survive in the ever-saltier water, and frequent foul smells and toxic dust are driving human residents away.

Join Asia and North America

The obvious place to link Asia and North America is at the Bering Strait, in between Russia's north-east corner and Alaska. At its narrowest point, the strait is just 82 kilometres across, and never more than 50 metres deep.

The idea of a bridge has been around since the 1890s. It would be the longest bridge over water, but not by a silly amount: the current record holder is the Qingdao-Haiwan bridge in China, which spans a 26-kilometre-wide stretch of water. But the Arctic conditions, especially the sea ice, pose a huge challenge. Oil drilling companies like Shell have struggled to even explore in the area.

That may be why Russia is more interested in a tunnel. In 2007, its government announced the TKM-World Link, a railway that would link Siberia to Alaska by way of a tunnel. A decade later, there is still no sign of the tunnel being dug, and relations between Russia and the US have soured. But perhaps China will take the lead: in 2014 the *Beijing Times* reported that engineers there are hatching plans for a high-speed railway that would run from China to the contiguous US, via Russia, the Bering Strait, Alaska and Canada.

It may not be a recipe for more harmonious relationships, however. Just over twenty years after the Channel Tunnel physically linked it to the Continent, the UK is in the process of breaking its political union with Europe.

Dam the Indian Ocean

Wherever there's a narrow bit of sea, someone has suggested installing concrete. The idea is usually to build a dam in a place where the water level on one side will drop because of evaporation. The resulting difference in height could be used to generate electricity.

There have been various proposals over the years but two stand out. In 2005, mega-engineering enthusiast Roelof Schuiling, a retired geochemist at Utrecht University in the Netherlands, suggested damming the Gulf in the Middle East where it opens into the Indian Ocean. At one point, the Strait of Hormuz, it narrows to just 39 kilometres across. The idea is not to do this any time soon, because it is an important shipping route for oil tankers. But when this trade declines, Schuiling says, damming the Indian Ocean and allowing the level of the Gulf to fall up to 35 metres could generate 2,500 megawatts of electricity.

There is an even bigger proposal out there: a dam across the Red Sea just before it joins the Indian Ocean, across the Bab-el-Mandeb Strait. That would require a dam wall 100 kilometres long, from Yemen in the north to either Eritrea or Djibouti in the south. Even Cathcart calls this 'a little more wild'. In 2007, he, Schuiling and their colleagues estimated it could generate around 50,000 megawatts of electricity.

These projects would lower local sea level and create more land. However, as with Atlantropa, they would cause sea level to rise even faster elsewhere. What's more, without any exchange with the Indian Ocean, the water in the seas would become steadily saltier, eventually destroying their entire ecosystems.

Creating land

Building artificial islands or peninsulas has become routine, with some astounding ones being made in Dubai, for example. But existing methods require deep quarries and deep pockets. Schuiling thinks there is a cheaper way to create land. He has shown that injecting sulphuric acid into limestone turns it into gypsum, causing it to swell to up to twice its original size. So where there is limestone close to the surface of the sea, new land could be created.

One such place is Adam's bridge, a narrow and shallow strip of shoals stretching for 35 kilometres between India and Sri Lanka. Schuiling thinks a land bridge could be created using his method for far less than the cost of a conventional bridge.

Relink the Pacific and Atlantic oceans

Destroying the Isthmus of Panama, the slender strip of land that joins North and South America, would reunite the Pacific and Atlantic oceans. Underground nuclear explosions would do the trick. With the land gone, the ocean current that once flowed around the equator would restart and, allegedly, stabilise the climate.

This idea is unlikely to be popular in Panama. What's more, some climate scientists think the closure of the gap 3 million years ago forced warm water in the tropical Atlantic to flow north, increasing humidity and snowfall in the Arctic and leading to the formation of the great northern ice sheets.

If so, nuking the isthmus would hasten the loss of the Greenland ice sheet.

Can geoengineering save us from climate change? Turn to page 156 to find out.

Could we save the world by going vegetarian?

? We're told that giving up meat is the most environmentally friendly way to feed yourself. But proving it turns out to be a tough nut to crack, finds Bob Holmes.

If you're a typical Westerner, you ate nearly 100 kilograms of meat last year. This was almost certainly the costliest part of your diet, especially in environmental terms. The clamour for people to eat less meat to save the planet is growing ever louder. 'Less meat = less heat', proclaimed Paul McCartney in the run-up to the 2009 conference on global warming in Copenhagen. Even *New Scientist* has recommended eating less meat as a way to reduce our environmental footprint.

If less is good, wouldn't none be better? You might think so. 'In the developed world, the most effective way to reduce the environmental impact of diet, on a personal basis, is to become vegetarian or vegan,' says Annette Pinner, chief executive of the Vegetarian Society in the UK. It seems like a no-brainer, but is it really that simple? To find out, let's imagine what would happen if the whole world decided to eliminate meat, milk and eggs from its diet, then trace the effects as they ripple throughout agriculture, the environment and society. The result may surprise you.

In 2015, the world consumed about 300 million tonnes of meat, 715 million tonnes of milk and 1.4 trillion eggs, according to the UN Food and Agriculture Organization (FAO). Environmentally speaking, this came at an enormous cost. All agriculture damages the environment – think of all those felled forests and ploughed-up prairies, all the irrigation water, manure, tractor fuel, pesticides and fertiliser. Agriculture produces more greenhouse gases than all methods of transport put together, and contributes to a host of other problems, from nitrogen pollution to soil erosion.

Livestock farming does the most damage. In part, that is because most livestock eat grain that could be used to feed people. As little as 10 per cent of that grain gets converted into meat, milk or eggs, so livestock amplify the environmental impact of farming by forcing us to grow more grain than we would otherwise need.

As a rough measure of how much more, consider that livestock consume about a third of the world's grain crop. So as a first approximation, a vegan world would need only two-thirds of the cropland used today. That's only part of the story, of course: meat and milk make up about 15 per cent of calories eaten by humans, so we would need to eat more grain to compensate for their loss. Altogether, switching to a vegan diet would reduce the amount of land used for crops by 21 per cent – about 3.4 million square kilometres, roughly the size of India.

Such a reduction would have a huge effect on the environmental impact of farming. Take nitrogen pollution, which can lead to eutrophication in lakes. As a small-scale illustration, environmental scientist Allison Leach of the University of Virginia in Charlottesville calculated that if everyone at her university cut out meat from their diet, it would reduce the university's nitrogen footprint – the amount of nitrogen released to the environment from all activities – by 27 per cent. This is largely because of reductions in fertiliser use and the amount of nitrogen leaching from manure. If everyone went a step farther and eliminated dairy products and eggs as well, Leach found that the university's nitrogen footprint would fall by 60 per cent.

It's not just in terms of nitrogen that livestock impact the environment. Global statistics are hard to come by, but in the US at least, livestock account for 55 per cent of soil erosion and 37 per cent of pesticide use. As well as that, half of all antibiotics manufactured are fed to livestock, often as part of their normal diet, a practice that is leading to antibiotic resistance in bacteria.

That's not all. Livestock are also a major source of greenhouse gases. Much of this comes in the form of methane – an especially potent greenhouse gas – produced by microbes in the guts of grazers such as cattle and sheep, and eventually belched out to the atmosphere. Livestock farming also accounts for a lot of carbon dioxide, mostly from forests being cut down for pasture, or when overgrazing and the resulting soil erosion causes a net loss of carbon from soils. When you add all this together, livestock account for a whopping 18 per cent of all greenhouse gas emissions as measured in CO_2 equivalents, according to *Livestock's Long Shadow*, a 2006 FAO report. Eliminating livestock would certainly make a big difference in efforts to control global warming.

Just how big a difference depends on what replaces the livestock and the land it grazes. Certainly, where pastures revert to forests – particularly in areas like the Amazon basin, for example, where 70 per cent of deforested land is now pasture – the regrowing forest will sequester huge amounts of carbon. The American plains, too, would accumulate carbon in their soil if grazing stopped. But in sub-Saharan Africa, any reduction in methane from domestic grazers is likely to be at least partially offset by increased emissions from wild grazers and termites, which compete with livestock for food. 'It's certainly worth someone spending some time to look at that,' says Philip Thornton, an agricultural systems scientist with the International Livestock Research Institute.

A meat-free world, then, would be greener in many ways: less cropland, more forest and, presumably, more biodiversity; lower greenhouse gas emissions; less agricultural pollution; less demand for fresh water – the list goes on. Clearly, if meat, milk and eggs were on trial for crimes against the environment, the prosecution would have an easy ride. And that says nothing of animal-welfare issues.

But wait. If everyone opted to give up meat there would be significant costs, too. It is true that most livestock today are fed grain that people could otherwise eat, but it doesn't have to be so. For most of human history, cattle, sheep and goats grazed on land that wasn't suitable for ploughing, and in doing so they converted inedible grass into edible meat and milk. Even today, a flock of sheep or goats can be the most efficient way to get food from marginal land. In a world where more than a billion people don't have enough to eat, taking such land out of production would only contribute to food insecurity. Moreover, for semi-arid or hilly land, modest levels of grazing may cause much less ecological damage than growing crops.

Even pigs and chickens, which lack the digestive machinery to eat grass, don't need grain. Instead they can subsist on leftovers and whatever they forage. 'Your household pig was your useful dustbin,' says Tara Garnett, who heads the Food Climate Research Network at Oxford University Centre for the Environment. 'You give your leftovers to the pigs, they deal with your rubbish, and you get meat.' Fed in this way, livestock represent a net gain of calories and protein in the human diet while dealing with some of the estimated 30 to

50 per cent of food that goes to waste – a benefit that a meat-free world would have to do without. Most pig and chicken farms are missing a trick here, since the animals eat commercial, grain-based feeds.

Another downside would be the disappearance of animal by-products. A meat-free world would have to replace the 11 million tonnes of leather and 2 million tonnes of wool that come from livestock farming every year. Not only that, many farmers would miss the manure, though the use of animal fertiliser is less important than it once was. 'Manure has become a minor source of nitrogen in all major agricultural countries. It's not unimportant, but it accounts for probably less than 15 per cent of total nitrogen,' says Vaclav Smil, an environmental scientist at the University of Manitoba in Winnipeg, Canada.

Even ardent vegetarians acknowledge that dairy products and even meat may be a good thing in poorer countries. 'Whilst there's no doubt that considerable reduction of meat consumption would have an environmental benefit, we do have to be careful about saying it would be the best solution if the whole world went vegetarian,' says Pinner. For as many as a billion of the world's poorest rural residents, an animal or two may represent their only realistic hope for a little extra income, and a little bit of animal protein can make a big difference to a marginal diet.

What if we decided on a vegetarian, rather than vegan, diet? After all, milk and eggs are very efficient ways of producing animal calories, second only to factory-reared broiler chickens. Unfortunately, an exclusively lacto-ovo livestock system simply doesn't work well in practice. 'It's difficult to switch to a no-meat but milk diet, because you cannot produce milk without meat,' says Helmut Haberl, a social ecologist at the Institute of Social Ecology in Vienna,

Austria. Dairy cows must calve every year to keep producing milk, and only half their offspring will be female. While many vegetarians see moral reasons not to kill and eat the males – or retired dairy cows – there is surely no practical reason to waste so much meat. Similar arguments apply to chickens kept for eggs.

So even though a meat-free world sounds good on paper, it is likely that a utopian future will still have some animal products in it. And we are talking meat, not just milk and eggs. The real questions, then, are how much meat do we want, and how will we produce it? The answers depend on how you approach the questions. The most straightforward way is to assume that the world will continue to demand ever more meat. That is certainly how things are going at the moment.

Under this scenario, the goal will have to be producing the most meat at the lowest environmental cost. That means fewer free-range cattle and sheep grazing in bucolic pastures and more animals, especially chickens, packed into feedlots or high-density enclosures. 'If you're going to keep some livestock systems, I think the ones you'll want to keep are the intensive ones,' says Walter Falcon, an agricultural economist at Stanford University in California.

That's because pasture grazing is inherently inefficient. Animals burn large amounts of energy roaming about the landscape feeding on relatively indigestible grasses. They grow more slowly than feedlot animals and, as a result, emit more methane over their lifetime. A beef cow in a US pasture, for example, emits 50 kilograms of methane per year, compared with just 26 kilograms in a feedlot, according to *Livestock's Long Shadow.*

But even a feedlot cow is a much less efficient meat producer than an industrial pig or chicken. While these eat a largely grain-based diet and thus compete directly with humans for food, they are relatively good at converting feed into flesh while producing little or no methane. This keeps their environmental cost down: a kilogram of industrial chicken meat represents greenhouse gas emissions equivalent to just 3.6 kilograms of CO_2; a kilogram of pork, 11.2 kilograms; and a kilogram of beef, 28.1 kilograms, according to an analysis by Bo Weidema of sustainable development consultancy 2.-0 LCA based in Aalborg, Denmark.

Of course, such intensive operations cause other problems as well, notably the disposal of large amounts of manure. In theory – and increasingly in practice – much of this manure could be used to generate biogas and subsequently electricity. If all US livestock manure were processed in this way, it could reduce greenhouse gas emissions by about 100 million tonnes annually, equivalent to 4 per cent of the emissions from electricity generation. With the right incentives, intensive livestock farms could cause much less environmental damage than they do today.

There is another alternative, though: treat livestock as part of the ecosystem. Garnett envisions returning animals to their original role as waste-disposal units, eating food leftovers and grazing on land not suitable for crops. 'In that context,' she says, 'methane emissions per animal will be higher, but overall emissions would be less because there would be fewer animals.'

Fewer animals means less meat, of course. Just how much less, no one really knows. As a first approximation, Garnett notes that about half of global meat production comes from intensive animal-only farms, and none of these would be allowed under the ecological approach. What is left would be those ranches where animals graze on marginal land and are not fed grain – about 10 per cent of the total today – and a larger number of mixed farms where the livestock feed off crop residues, milling wastes and other leftovers.

Such a future would require a major adjustment in food preferences. People would need to eat less meat, especially in the meat-hungry West. Not only that, but we would also have to change the kind of meat we eat. 'You are not going to get your fat, heavy-breasted chickens by feeding them household scraps and letting them peck for worms. You are going to get a much scrawnier animal,' says Garnett.

Would people really accept pricey free-range beef and scrawny barnyard chickens perhaps once or twice a week? Certainly most do not today, opting for price and abundance over environmental impact. But change happens. Given the deforestation, soil erosion, water pollution and greenhouse gas emissions that will result if worldwide meat production continues to rise, some people are already choosing to eat less meat. And the message is definitely less, not none. For best results, meat should be medium-rare.

Is there an alternative to countries?

? **Nation-states cause some of our biggest problems, from civil war to climate inaction. Science suggests there are better ways to run a planet, says Debora MacKenzie.**

Try, for a moment, to envisage a world without countries. Imagine a map not divided into neat, coloured patches, each with clear borders, governments, laws. Try to describe anything our society does – trade, travel, science, sport, maintaining peace and security – without mentioning countries. Try to describe yourself: you have a right to at least one

nationality, and the right to change it, but not the right to have none.

Those coloured patches on the map may be democracies, dictatorships or too chaotic to be either, but virtually all claim to be one thing: a nation-state, the sovereign territory of a 'people' or nation who are entitled to self-determination within a self-governing state. So says the United Nations, which now numbers 193 of them.

And more and more peoples want their own state, from Scots voting for independence to jihadis declaring a new state in the Middle East. Many of the big news stories of the day, from conflicts in Gaza and Ukraine to rows over immigration and membership of the European Union, are linked to nation-states in some way.

Even as our economies globalise, nation-states remain the planet's premier political institution. Large votes for nationalist parties in recent EU elections prove nationalism remains alive – even as the EU tries to transcend it.

Yet there is a growing feeling among economists, political scientists and even national governments that the nation-state is not necessarily the best scale on which to run our affairs. We must manage vital matters like food supply and climate on a global scale, yet national agendas repeatedly trump the global good. At a smaller scale, city and regional administrations often seem to serve people better than national governments.

How, then, should we organise ourselves? Is the nation-state a natural, inevitable institution? Or is it a dangerous anachronism in a globalised world? These are not normally scientific questions – but that is changing. Complexity theorists, social scientists and historians are addressing them using new techniques, and the answers are not always what you might expect. Far from timeless, the nation-state is a recent phenomenon. And as complexity keeps rising, it is already mutating into novel political structures. Get set for neo-medievalism.

Before the late eighteenth century, there were no real nation-states, says John Breuilly of the London School of Economics. If you travelled across Europe, no one asked for your passport at borders; neither passports nor borders as we know them existed. People had ethnic and cultural identities, but these didn't really define the political entity they lived in.

That goes back to the anthropology, and psychology, of humanity's earliest politics. We started as wandering, extended families, then formed larger bands of hunter-gatherers, and then, around 10,000 years ago, settled in farming villages. Such alliances had adaptive advantages, as people cooperated to feed and defend themselves.

But they also had limits. Robin Dunbar has shown that one individual can keep track of social interactions linking no more than around 150 people. Evidence for that includes studies of villages and army units through history, and the average tally of Facebook friends.

But there was one important reason to have more friends than that: war. 'In small-scale societies, between 10 and 60 per cent of male deaths are attributable to warfare,' says Peter Turchin of the University of Connecticut at Storrs. More allies meant a higher chance of survival.

Turchin has found that ancient Eurasian empires grew largest where fighting was fiercest, suggesting war was a major factor in political enlargement. Archaeologist Ian Morris of Stanford University in California reasons that as populations grew, people could no longer find empty lands where they could escape foes. The losers of battles were simply absorbed into the enemy's domain – so domains grew bigger.

How did they get past Dunbar's number? Humanity's universal answer was the invention of hierarchy. Several villages allied themselves under a chief; several chiefdoms banded together under a higher chief. To grow, these alliances added more villages, and if necessary more layers of hierarchy.

Hierarchies meant leaders could coordinate large groups without anyone having to keep personal track of more than 150 people. In addition to their immediate circle, an individual interacted with one person from a higher level in the hierarchy, and typically eight people from lower levels, says Turchin.

These alliances continued to enlarge and increase in complexity in order to perform more kinds of collective actions, says Yaneer Bar-Yam of the New England Complex Systems Institute in Cambridge, Massachusetts. For a society to survive, its collective behaviour must be as complex as the challenges it faces – including competition from neighbours. If one group adopted a hierarchical society, its competitors also had to. Hierarchies spread and social complexity grew.

Larger hierarchies not only won more wars but also fed more people through economies of scale, which enabled technical and social innovations such as irrigation, food storage, record-keeping and a unifying religion. Cities, kingdoms and empires followed.

But these were not nation-states. A conquered city or region could be subsumed into an empire regardless of its inhabitants' 'national' identity. 'The view of the state as a necessary framework for politics, as old as civilisation itself, does not stand up to scrutiny,' says historian Andreas Osiander of the University of Leipzig in Germany.

One key point is that agrarian societies required little actual governing. Nine people in ten were peasants who had to farm or starve, so were largely self-organising. Government intervened to take its cut, enforce basic criminal law and keep the peace within its undisputed territories. Otherwise its main role was to fight to keep those territories, or acquire more.

Even quite late on, rulers spent little time governing, says Osiander. In the seventeenth century Louis XIV of France had half a million troops fighting foreign wars but only 2,000 keeping order at home. In the eighteenth century, the Dutch and Swiss needed no central government at all. Many eastern European immigrants arriving in the US in the nineteenth century could say what village they came from, but not what country: it didn't matter to them.

Before the modern era, says Breuilly, people defined themselves 'vertically' by who their rulers were. There was little horizontal interaction between peasants beyond local markets. Whoever else the king ruled over, and whether those people were anything like oneself, was largely irrelevant.

Such systems are very different from today's states, which have well-defined boundaries filled with citizens. In a system of vertical loyalties, says Breuilly, power peaks where the overlord lives and peters out in frontier territories that shade into neighbouring regions. Ancient empires are coloured on modern maps as if they had firm borders, but they didn't. Moreover, people and territories often came under different jurisdictions for different purposes.

Such loose control, says Bar-Yam, meant pre-modern political units were only capable of scaling up a few simple actions such as growing food, fighting battles, collecting tribute and keeping order. Some, like the Roman Empire, did this on a very large scale. But complexity – the different actions society could collectively perform – was relatively low.

Complexity was limited by the energy a society could

harness. For most of history that essentially meant human and animal labour. In the late Middle Ages, Europe harnessed more, especially water power. This boosted social complexity – trade increased, for example – requiring more government. A decentralised feudal system gave way to centralised monarchies with more power.

But these were still not nation-states. Monarchies were defined by who ruled them, and rulers were defined by mutual recognition – or its converse, near-constant warfare. In Europe, however, as trade grew, monarchs discovered they could get more power from wealth than war.

In 1648, Europe's Peace of Westphalia ended centuries of war by declaring existing kingdoms, empires and other polities 'sovereign': none was to interfere in the internal affairs of others. This was a step towards modern states – but these sovereign entities were still not defined by their peoples' national identities. International law is said to date from the Westphalia treaty, yet the word 'international' was not coined until 132 years later.

By then Europe had hit the tipping point of the industrial revolution. Harnessing vastly more energy from coal meant that complex behaviours performed by individuals, such as weaving, could be amplified, says Bar-Yam, producing much more complex collective behaviours.

This demanded a different kind of government. In 1776 and 1789, revolutions in the US and France created the first nation-states, defined by the national identity of their citizens rather than the bloodlines of their rulers. According to one landmark history of the period, says Breuilly, 'in 1800 almost nobody in France thought of themselves as French. By 1900 they all did.' For various reasons, people in England had an earlier sense of 'Englishness', he says, but it was not expressed as a nationalist ideology.

By 1918, with the dismemberment of Europe's last multinational empires such as the Habsburgs in the first world war, European state boundaries had been redrawn largely along cultural and linguistic lines. In Europe at least, the nation-state was the new norm.

Part of the reason was a pragmatic adaptation of the scale of political control required to run an industrial economy. Unlike farming, industry needs steel, coal and other resources which are not uniformly distributed, so many micro-states were no longer viable. Meanwhile, empires became unwieldy as they industrialised and needed more actual governing. So in nineteenth-century Europe, micro-states fused and empires split.

These new nation-states were justified not merely as economically efficient, but as the fulfilment of their inhabitants' national destiny. A succession of historians has nonetheless concluded that it was the states that defined their respective nations, and not the other way around.

France, for example, was not the natural expression of a pre-existing French nation. At the revolution in 1789, half its residents did not speak French. In 1860, when Italy unified, only 2.5 per cent of residents regularly spoke standard Italian. Its leaders spoke French to each other. One famously said that, having created Italy, they now had to create Italians – a process many feel is still taking place.

Sociologist Siniša Maleševic of University College Dublin in Ireland believes that this 'nation building' was a key step in the evolution of modern nation-states. It required the creation of an ideology of nationalism that emotionally equated the nation with people's Dunbar circle of family and friends.

That in turn relied heavily on mass communication technologies. In an influential analysis, Benedict Anderson of

Cornell University in New York described nations as 'imagined' communities: they far outnumber our immediate circle and we will never meet them all, yet people will die for their nation as they would for their family.

Such nationalist feelings, he argued, arose after mass-market books standardised vernaculars and created linguistic communities. Newspapers allowed people to learn about events of common concern, creating a large 'horizontal' community that was previously impossible. National identity was also deliberately fostered by state-funded mass education.

The key factor driving this ideological process, Maleševic says, was an underlying structural one: the development of far-reaching bureaucracies needed to run complex industrialised societies. For example, says Breuilly, in the 1880s Prussia became the first government to pay unemployment benefits. At first they were paid only in a worker's native village, where identification was not a problem. As people migrated for work, benefits were made available anywhere in Prussia. 'It wasn't until then that they had to establish who a Prussian was,' he says, and they needed bureaucracy to do it. Citizenship papers, censuses and policed borders followed.

That meant hierarchical control structures ballooned, with more layers of middle management. Such bureaucracy was what really brought people together in nation-sized units, argues Maleševic. But not by design: it emerged out of the behaviour of complex hierarchical systems. As people do more kinds of activities, says Bar-Yam, the control structure of their society inevitably becomes denser.

In the emerging nation-state, that translates into more bureaucrats per head of population. Being tied into such close bureaucratic control also encouraged people to feel personal ties with the state, especially as ties to church and village declined. As governments exerted greater control, people got more rights, such as voting, in return. For the first time, people felt the state was theirs.

Once Europe had established the nation-state model and prospered, says Breuilly, everyone wanted to follow suit. In fact it's hard now to imagine that there could be another way. But is a structure that grew spontaneously out of the complexity of the industrial revolution really the best way to manage our affairs?

According to Brian Slattery of York University in Toronto, Canada, nation-states still thrive on a widely held belief that 'the world is naturally made of distinct, homogeneous national or tribal groups which occupy separate portions of the globe, and claim most people's primary allegiance'. But anthropological research does not bear that out, he says. Even in tribal societies, ethnic and cultural pluralism has always been widespread. Multilingualism is common, cultures shade into each other, and language and cultural groups are not congruent.

Moreover, people always have a sense of belonging to numerous different groups based on region, culture, background and more. 'The claim that a person's identity and well-being is tied in a central way to the well-being of the national group is wrong as a simple matter of historical fact,' says Slattery. Perhaps it is no wonder, then, that the nation-state model fails so often: since 1960 there have been more than 180 civil wars worldwide.

Such conflicts are often blamed on ethnic or sectarian tensions. Failed states are typically riven by violence along such lines. According to the idea that nation-states should contain only one nation, such failures have often been blamed on the colonial legacy of bundling together many peoples within unnatural boundaries.

But for every Syria or Iraq there is a Singapore, Malaysia or Tanzania, getting along OK despite having several 'national' groups. Immigrant states in Australia and the Americas, meanwhile, forged single nations out of massive initial diversity.

What makes the difference? It turns out that while ethnicity and language are important, what really matters is bureaucracy. This is clear in the varying fates of the independent states that emerged as Europe's overseas empires fell apart after the second world war.

According to the mythology of nationalism, all they needed was a territory, a flag, a national government and UN recognition. In fact what they really needed was complex bureaucracy.

Some former colonies that had one became stable democracies, notably India. Others did not, especially those such as the former Belgian Congo, whose colonial rulers had merely extracted resources. Many of these became dictatorships, which require a much simpler bureaucracy than democracies.

Dictatorships exacerbate ethnic strife because their institutions do not promote citizens' identification with the nation. In such situations, people fall back on trusted alliances based on kinship, which readily elicit Dunbar-like loyalties. Insecure governments allied to ethnic groups favour their own, while grievances among the disfavoured groups grow – and the resulting conflict can be fierce.

Recent research confirms that the problem is not ethnic diversity itself, but not enough official inclusiveness. Countries with little historic ethnic diversity are now having to learn that on the fly, as people migrate to find jobs within a globalised economy.

How that pans out may depend on whether people self-segregate. Humans like being around people like themselves, and ethnic enclaves can be the result. Jennifer Neal of Michigan State University in East Lansing has used agent-based modelling to look at the effect of this in city neighbourhoods. Her work suggests that enclaves promote social cohesion, but at the cost of decreasing tolerance between groups. Small enclaves in close proximity may be the solution.

But at what scale? Bar-Yam says communities where people are well mixed – such as in peaceable Singapore, where enclaves are actively discouraged – tend not to have ethnic strife. Larger enclaves can also foster stability. Using mathematical models to correlate the size of enclaves with the incidences of ethnic strife in India, Switzerland and the former Yugoslavia, he found that enclaves 56 kilometres or more wide make for peaceful coexistence – especially if they are separated by natural geographical barriers.

Switzerland's twenty-six cantons, for example, which have different languages and religions, meet Bar-Yam's spatial stability test – except one. A French-speaking enclave in German-speaking Berne experienced the only major unrest in recent Swiss history. It was resolved by making it a separate canton, Jura, which meets the criteria.

Again, though, ethnicity and language are only part of the story. Lars-Erik Cederman of the Swiss Federal Institute of Technology in Zurich argues that Swiss cantons have achieved peace not by geographical adjustment of frontiers, but by political arrangements giving cantons considerable autonomy and a part in collective decisions.

Similarly, using a recently compiled database to analyse civil wars since 1960, Cederman finds that strife is indeed more likely in countries that are more ethnically diverse. But careful analysis confirms that trouble arises not from diversity alone, but when certain groups are systematically excluded from power.

Governments with ethnicity-based politics were especially vulnerable. The US set up just such a government in Iraq after the 2003 invasion. Exclusion of Sunni by Shiites led to insurgents declaring a Sunni state in occupied territory in Iraq and Syria. True to nation-state mythology, it rejects the colonial boundaries of Iraq and Syria, as they force dissimilar 'nations' together.

Yet the solution cannot be imposing ethnic uniformity. Historically, so-called ethnic cleansing has been uniquely bloody, and 'national' uniformity is no guarantee of harmony. In any case, there is no good definition of an ethnic group. Many people's ethnicities are mixed and change with the political weather: the numbers who claimed to be German in the Czech Sudetenland territory annexed by Hitler changed dramatically before and after the war. Russian claims to Russian-speakers in eastern Ukraine now may be equally flimsy.

The research of both Bar-Yam and Cederman suggests one answer to diversity within nation-states: devolve power to local communities, as multicultural states such as Belgium and Canada have done. 'We need a conception of the state as a place where multiple affiliations and languages and religions may be safe and flourish,' says Slattery. 'That is the ideal Tanzania has embraced and it seems to be working reasonably well.' Tanzania has more than 120 ethnic groups and about a hundred languages.

In the end, what may matter more than ethnicity, language or religion is economic scale. The scale needed to prosper may have changed with technology – tiny Estonia is a high-tech winner – but a small state may still not pack enough economic power to compete.

That is one reason why Estonia is such an enthusiastic member of the European Union. After the devastating wars in the twentieth century, European countries tried to prevent further war by integrating their basic industries. That project, which became the European Union, now primarily offers member states profitable economies of scale, through manufacturing and selling in the world's largest single market.

What the EU fails to inspire is nationalist-style allegiance – which Maleševic thinks nowadays relies on the 'banal' nationalism of sport, anthems, TV news programmes, even song contests. That means Europeans' allegiances are no longer identified with the political unit that handles much of their government.

Ironically, says Jan Zielonka of the University of Oxford, the EU has saved Europe's nation-states, which are now too small to compete individually. The call by nationalist parties to 'take back power from Brussels', he argues, would lead to weaker countries, not stronger ones.

He sees a different problem. Nation-states grew out of the complex hierarchies of the industrial revolution. The EU adds another layer of hierarchy – but without enough underlying integration to wield decisive power. It lacks both of Maleševic's necessary conditions: nationalist ideology and pervasive integrating bureaucracy.

Even so, the EU may point the way to what a post-nation-state world will look like. Zielonka agrees that further integration of Europe's governing systems is needed as economies become more interdependent. But he says Europe's often-paralysed hierarchy cannot achieve this. Instead he sees the replacement of hierarchy by networks of cities, regions and even non-governmental organisations. Sound familiar? Proponents call it neo-medievalism.

'The future structure and exercise of political power will resemble the medieval model more than the Westphalian one,' Zielonka says. 'The latter is about concentration of power, sovereignty and clear-cut identity.' Neo-medievalism, on the other hand, means overlapping authorities, divided sovereignty, multiple identities and governing institutions, and fuzzy borders.

Anne-Marie Slaughter of Princeton University, a former US assistant secretary of state, also sees hierarchies giving way to global networks primarily of experts and bureaucrats from nation-states. For example, governments now work more through flexible networks such as the G7 (or 8, or 20) to manage global problems than through the UN hierarchy.

Ian Goldin, professor of Globalisation and Development at the University of Oxford, thinks such networks must emerge. He believes existing institutions such as UN agencies and the World Bank are structurally unable to deal with problems that emerge from global interrelatedness, such as economic instability, pandemics, climate change and cybersecurity – partly because they are hierarchies of member states which themselves cannot deal with these global problems. He quotes Slaughter: 'Networked problems require a networked response.'

Again, the underlying behaviour of systems and the limits of the human brain explain why. Bar-Yam notes that in any hierarchy, the person at the top has to be able to get their head around the whole system. When systems are too complex for one human mind to grasp, he argues that they must evolve from hierarchies into networks where no one person is in charge.

Where does this leave nation-states? 'They remain the main containers of power in the world,' says Breuilly. And we need their power to maintain the personal security that has permitted human violence to decline to all-time lows.

Moreover, says Dani Rodrik of Princeton's Institute for Advanced Study, the very globalised economy that is allowing these networks to emerge needs something or somebody to write and enforce the rules. Nation-states are currently the only entities powerful enough to do this.

Yet their limitations are clear, both in solving global problems and resolving local conflicts. One solution may be to pay more attention to the scale of government. Known as subsidiarity, this is a basic principle of the EU: the idea that government should act at the level where it is most effective, with local government for local problems and higher powers at higher scales. There is empirical evidence that it works: social and ecological systems can be better governed when their users self-organise than when they are run by outside leaders.

However, it is hard to see how our political system can evolve coherently in that direction. Nation-states could get in the way of both devolution to local control and networking to achieve global goals. With climate change, it is arguable that they already have.

There is an alternative to evolving towards a globalised world of interlocking networks, neo-medieval or not, and that is collapse. 'Most hierarchical systems tend to become top-heavy, expensive and incapable of responding to change,' says Marten Scheffer of Wageningen University in the Netherlands. 'The resulting tension may be released through partial collapse.' For nation-states, that could mean anything from the renewed pre-eminence of cities to Iraq-style anarchy. An uncertain prospect, but there is an upside. Collapse, say some, is the creative destruction that allows new structures to emerge.

Like it or not, our societies may already be undergoing this transition. We cannot yet imagine there are no countries. But recognising that they were temporary solutions to specific historical situations can only help us manage a transition to whatever we need next. Whether or not our nations endure, the structures through which we govern our affairs are due for a change. Time to start imagining.

5 What Happens Next?

Predicting the future can be a risky business, unless you're a many-worlds theorist. Which of a thousand possible future timelines might we find ourselves in half a century from now? All of them, of course. Does that make forecasting a worthless endeavour? Not at all: some futures are still more likely than others.

In 2016, *New Scientist* celebrated its diamond jubilee – sixty years of reporting what's new in science and why it matters. To mark the occasion, we produced a special issue exploring what the world might look like on our 120th anniversary. Will we have found a solution to climate change? Or settled on Mars? Would robots rule Earth?

'We do not simply insist that we reside in the best of all possible worlds,' editor Sumit Paul-Choudhury wrote at the time. 'We think we have to make it so.' In this chapter, we think about the future, specifically where we might be in 2076. Whether these scenarios come to pass – well, that depends on which timeline you decide to pursue.

Could the climate be controlled?

? Epic geoengineering megaprojects could save us from warming. But Catherine Brahic discovers this high-stakes gamble risks unleashing a catastrophe.

It's 2076 and the skies are looking decidedly milky. On windy plains and in parts of the seas that have been turned over to wind farms, a different kind of tower has been built alongside the turbines. They suck carbon dioxide out of the atmosphere. Vast parcels of land have been given over to forest. Trees are grown, harvested and burned for energy in power plants that don't let CO_2 escape to the atmosphere. Instead, emissions are captured and pumped into underground storage reservoirs. Ships dump powdered minerals into the water to soak up CO_2 and reduce ocean acidification.

All these technologies are a desperate rearguard action to reverse more than two centuries of greenhouse gas emissions. But they are not entirely up to the task and, anyway, we are still emitting greenhouse gases. So, 10 to 18 kilometres up in the atmosphere, a fine spray of particles shields Earth from the sun and keeps us cool. It's what is making the skies that little bit whiter.

'I think it's very likely that in sixty years we'll be using both technologies,' says John Shepherd of the University of Southampton, UK. He is referring to the two flavours of geoengineering: sucking CO_2 out of the air and deploying a sunshade to bounce some of the sun's rays back out into space.

Like many climate scientists, Shepherd thinks climate talks are going too slowly. Even if industrial emissions were to drop rapidly – a big if – some sectors pose an intractable problem. We have no real replacement for aeroplane fuel and feeding people demands intensive agriculture, which accounts for a quarter of global emissions. That is why we will have to suck CO_2 out of the air. And because that is a long way off, we will probably also have to rely on 'solar radiation management'.

The most studied version is spraying fine particles of sulphate into the stratosphere, yet its consequences are still poorly understood. Computer models suggest there will be winners and losers. While a sunshade could lower global average temperatures to pre-industrial levels, there would be regional differences. Northern Europe, Canada, Siberia and the poles would remain warmer than they were, and temperatures over the oceans would be cooler.

Global warming is predicted to make wet regions wetter and dry ones drier. Models suggest a sunshade would rectify this, but, again, not in a uniform way. Tropical regions that depend on seasonal rains could suffer most, with monsoons drying up.

Shepherd fears all this will feed into international disputes. He envisages some kind of global council where governments lobby for a climate that meets their needs. Some might prefer a slightly warmer temperature, for tourism or agriculture. But nations whose coral reefs draw in visitors will probably want more CO_2-sucking technologies to counter ocean acidification and bleaching.

There's a final, bitter twist. What goes up must come down, and so a sunshade would have to be continually replenished. If something were to happen – say, a complete breakdown of some international geoengineering agreement – and we stopped spraying sulphates, the consequences would be catastrophic. Within a decade or two, temperatures would soar to where they would have been without the sunshade. Warm regions would fry, and who knows what tipping points we would fly past. The consequences are barely worth thinking about.

Will human-made life forms roam the planet?

? Synthetic life is well within our capabilities, says Bob Holmes – but creating a free-living, independently evolving life form also comes with huge risks.

Life arose on Earth almost as soon as the planet had cooled enough to be habitable – and as far as we know, it has never arisen again in the 4 billion years since. That long dry spell may end within the next few years, though, as researchers near the goal of making life from scratch in the lab.

Already, geneticists have synthesised a bespoke genome and inserted it into a bacterium. They have also altered the genetic code of other bacteria to get them to use new, non-natural building blocks to make proteins. But all these efforts start with a living organism and merely modify it.

A more ambitious effort starts with nonliving, chemical ingredients – sometimes familiar nucleic acids and lipids, but sometimes radically different structures such as self-assembling metal oxides. The researchers aim to coax these chemicals across the Darwinian threshold where they begin to replicate themselves heritably and evolve – the key criteria for calling the system alive. If this can be achieved, the implications would be enormous.

Most fundamentally, synthetic life would complete the philosophical break – one that Darwin started – from a creation-centred view of the living world. 'It'll prove pretty decisively that life is nothing more than a complicated chemical system,' says Mark Bedau, a philosopher of science at Reed College in Portland, Oregon. Most scientists already think this way, of course, but synthetic life would make the point in a way the wider world could not ignore. Moreover, creating it in the lab would prove that the origin of life is a relatively low hurdle, increasing the odds that we might find life elsewhere in the solar system.

A second genesis would also give biologists an independent point of comparison to understand what makes life tick. And because we made it, we would be able to modify it, changing the ingredients to learn which features are truly essential.

The stuff we will end up calling 'natural life' is so encumbered with billions of years of evolutionary baggage that it's impossible to distinguish between what is truly essential for life and what has become essential for our particular sort, says Steven Benner from the Foundation for Applied Molecular Evolution in Florida. Newly created life would give experimentalists a cleaner system for testing life's needs.

Practical payoffs are likely to be farther in the future. Any new life form would be so feeble at first that it couldn't survive without coddling in the lab, so biotechnologists who want to produce particular molecules or degrade toxic waste, for example, will have better success modifying natural life. In the long run, however, artificial life might grow robust enough to thrive on its own. If so, it would allow biotechnologists to escape the constraints of natural life to accomplish new goals. 'We can explore all sorts of possible payoffs,' says Lee Cronin at the University of Glasgow, UK.

But those benefits bring risks, too. A free-living, independently evolving life form is, by definition, no longer entirely predictable or controllable. Biotechnologists will need to design effective 'kill switches' in case the new life becomes pathogenic or harmful in other ways, and policy-makers and ethicists will need to work out when and how to trigger them. The public may try to stymie the whole enterprise, amid the usual accusations of playing God. Discussions on the implications of synthetic life need to start soon. 'Within a short time, this could be a serious issue,' says Bedau.

What if we don't need bodies?

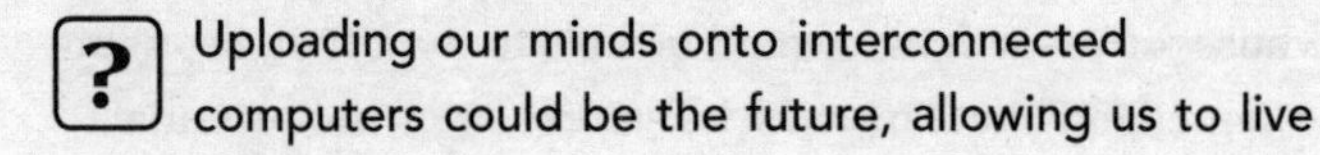

Uploading our minds onto interconnected computers could be the future, allowing us to live

in any version of the multiverse we please. But cutting ties with our animal roots would raise ethical questions for which we don't yet have answers, say Anil Ananthaswamy and MacGregor Campbell.

Minds result from bodies, but that link can be compromised. 'If I severed my spinal cord at the neck, I'd get no inputs from most of my body,' says Michael Graziano, a neuroscientist at Princeton University. 'But I'm still a person, I still have experience, I can still think.'

What if we could separate mind from body entirely? Many now believe that we will transfer our minds onto computers, whether in a matter of decades or hundreds of years. 'I would say that it's not only possible, it's inevitable,' says Graziano.

What would life as an upload be like? We'd still need outside stimulation. Cut off entirely, a brain would suffer sensory deprivation, says Anders Sandberg at the University of Oxford. 'It's going to fall asleep, then hallucinate and probably gently go mad. You need to give it a way of interacting with the world, although it doesn't have to be the real world.'

Being able to transfer minds into a computer would change how we valued a life. Having multiple backup copies of ourselves might make life less precious. 'You kill one of them, so what?' says Graziano. 'There's a whole bunch more.' Murder may no longer be a heinous crime when we can resurrect the dead. The same goes for victims of freak accidents, says Sandberg. Just boot up the last save and the only thing lost may be a few recent memories.

And if you think we are living in a hyperconnected world today, think again. Artificial brains would give connectivity a whole different meaning. 'Forget texting, you can as good as stick a USB port in your head and communicate directly with somebody else,' says Graziano. 'Now, we get into a totally different network of minds that doesn't resemble anything we know.' Instead of inferring what is going on in someone else's head, we could share thoughts as we share digital files. This 'noosphere' could enable true global consciousness – but it might also obliterate the individual, transforming our existential landscape for ever.

But even if we could establish connections with the required fidelity, we will have a translation problem. 'My mind doesn't work like your mind,' Sandberg says. Creating software that can translate different mental representations of various concepts might be as challenging as creating human-level artificial intelligence.

There may be a workaround. The brain's plasticity allows it to incorporate and interpret new sensory information. Sandberg thinks that with the right technology we might train our neocortices, the regions of our brains responsible for consciousness, to adapt to more complex signals coming from other brains, rather than from simple sensors.

What might life in the hive mind be like? Acting as part of a group can be joyous and fulfilling, and the larger the group, the greater the benefit. So joining a global noosphere could be a profound and ecstatic experience. We might all share the joy of holding a newborn baby, multiplied by the 350,000 born around the world every day, say, or marvel at how quickly billions of coordinated hands can fix the environment.

But there is a dark side. 'If technology makes it easy for the good ideas to spread, it can also make it easy for the stupid ideas,' says Sandberg. False accusations, for instance, could rage through our shared consciousness like wildfire, supercharging the worst that mob rule has to offer.

Advanced neural filters that automatically block the

most dangerous thoughts might prevent the worst-case scenarios, says Sandberg. The same goes for securing our minds against brain-hackers seeking to influence or even directly control our thoughts and desires. But such filters would have to assess the content of neural signals to understand human thought, a staggeringly complex task to say the least.

A big concern will be who controls the computers that run the brain simulations. In principle, those running the machines could make copies of you, says Sandberg. 'They could run you on a secret system with no connection to the internet and force you to do a lot of stuff.' We will need to drastically improve our software security in general, he says. 'If it was a world where anybody could be hacked or copied at any time by unknown parties, it'd be a bit too scary to live in.'

Technical issues will abound. We may make mistakes during the upload process, and have warped brains sitting in computers. 'Do you want a library of bad copies that you have some weird obligation to?' says Sandberg. The ethical conundrums become even more complicated if we assume your original brain and body stick around. Would virtual brains have equal moral and legal status?

Such questions are unlikely to be resolved in a hurry. 'In many ways, we would become post-human,' says Sandberg. 'We'd have made the leap from being part of the animal kingdom to going into an entirely new kingdom, and we don't know what to call it yet.'

If all such hurdles are overcome, the hive mind might operate at different scales, says Sandberg. Our local individual experience would still be ours, as long as the security measures hold up, but we might choose to switch viewpoints, as in a video game. And we might modulate signals coming from higher levels – family, city, regional and global – so that we experience them as our own preferences or even gut feelings.

However, as in the early days of the internet, you will probably have to get used to buffering. Nerve impulses move more slowly than the signals between computers. Multiply the inevitable lag by billions of brains, and the hive mind might feel positively indecisive.

Even in the deepest future, the speed of light will impose limits on what a hive mind can do, says Sandberg. 'A universe-scale hive mind might take billions of years to think a single thought.'

Is consciousness just a state of matter?

? If consciousness is nothing more than an arrangement of atoms, what's to stop any complex system gaining sentience? The atomic arrangement that calls itself Max Tegmark asks if consciousness is the fourth state of matter.

Why are you conscious right now? Specifically, why are you having a subjective experience of reading these words, seeing colours and hearing sounds, while the inanimate objects around you presumably aren't having any subjective experience at all?

Different people mean different things by 'consciousness', including awareness of environment or self. I am asking the more basic question of why you experience anything at all, which is the essence of what philosopher David Chalmers has coined 'the hard problem' of consciousness.

A traditional answer to this problem is dualism – that living entities differ from inanimate ones because they contain some non-physical element such as an 'anima' or 'soul'. Support for dualism among scientists has gradually dwindled. To understand why, consider that your body is made up of about 10^{29} quarks and electrons, which as far as we can tell move according to simple physical laws. Imagine a future technology able to track all of your particles: if they were found to obey the laws of physics exactly, then your purported soul is having no effect on your particles, so your conscious mind and its ability to control your movements would have nothing to do with a soul.

If your particles were instead found not to obey the known laws of physics because they were being pushed around by your soul, then we could treat the soul as just another physical entity able to exert forces on particles, and study what physical laws it obeys.

Let us therefore explore the other option, known as physicalism: that consciousness is a process that can occur in certain physical systems. This raises a fascinating question: why are some physical entities conscious, while others are not? If we consider the most general state of matter that experiences consciousness – let's call it 'perceptronium' – then what special properties does it have that we could in principle measure in a lab? What are these physical correlates of consciousness? Parts of your brain clearly have these properties right now, as well as while you were dreaming last night, but not while you were in deep sleep.

Imagine all the food you have eaten in your life and consider that you are simply some of that food, rearranged. This shows that your consciousness isn't simply due to the atoms you ate, but depends on the complex patterns into which these atoms are arranged. If you can also imagine conscious entities, say aliens or future superintelligent robots, made out of different types of atoms then this suggests that consciousness is an 'emergent phenomenon' whose complex behaviour emerges from many simple interactions. In a similar spirit, generations of physicists and chemists have studied what happens when you group together vast numbers of atoms, finding that their collective behaviour depends on the patterns in which they are arranged. For instance, the key difference between a solid, a liquid and a gas lies not in the types of atoms, but in their arrangement. Boiling or freezing a liquid simply rearranges its atoms.

My hope is that we will ultimately be able to understand perceptronium as yet another state of matter. Just as there are many types of liquids, there are many types of consciousness. However, this should not preclude us from identifying, quantifying, modelling and understanding the characteristic properties shared by all liquid forms of matter, or all conscious forms of matter. Take waves, for example, which are substrate-independent in the sense that they can occur in all liquids, regardless of their atomic composition. Like consciousness, waves are emergent phenomena in the sense that they take on a life of their own: a wave can traverse a lake while the individual water molecules merely bob up and down, and the motion of the wave can be described by a mathematical equation that doesn't care what the wave is made of.

Something analogous happens in computing. Alan Turing famously proved that all sufficiently advanced computers can simulate one another, so a video-game character in her virtual world would have no way of knowing whether her computational substrate ('computronium') was a Mac or a PC, or what types of atoms the hardware was made of. All that would matter is abstract information processing. If this created character were complex enough to be conscious, as in the film *The Matrix*, then what properties would this information processing need to have?

I have long contended that consciousness is the way information feels when processed in certain complex ways. The neuroscientist Giulio Tononi has made this idea more specific and useful, making the compelling argument that for an information processing system to be conscious, its information must be integrated into a unified whole. In other words, it must be impossible to decompose the system into nearly independent parts – otherwise these parts would feel like two separate conscious entities. Tononi and his collaborators have incorporated this idea into an elaborate mathematical formalism known as integrated information theory (IIT).

IIT has generated significant interest in the neuroscience community, because it offers answers to many intriguing questions. For example, why do some information processing systems in our brains appear to be unconscious? Based on extensive research correlating brain measurements with subjectively reported experience, neuroscientist Christof Koch and others have concluded that the cerebellum – a brain area whose roles include motor control – is not conscious, but is an unconscious information processor that helps other parts of the brain with certain computational tasks.

The IIT explanation for this is that the cerebellum is mainly a collection of 'feed-forward' neural networks in which information flows like water down a river, and each neuron affects mostly those downstream. If there is no feedback, there is no integration and hence no consciousness. The same would apply to Google's feed-forward artificial neural network that processed millions of YouTube video frames to determine whether they contained cats. In contrast, the brain systems linked to consciousness are strongly integrated, with all parts able to affect one another.

IIT thus offers an answer to the question of whether a superintelligent computer would be conscious: it depends. A part of its information processing system that is highly integrated will indeed be conscious. However, IIT research has shown that for many integrated systems, one can design a functionally equivalent feed-forward system that will be unconscious. This means that so-called 'p-zombies' can, in principle, exist: systems that behave like a human and pass the Turing test for machine intelligence, yet lack any conscious experience whatsoever. Many current 'deep learning' AI systems are of this p-zombie type. Fortunately, integrated systems such as those in our brains typically require far fewer computational resources than their feed-forward 'zombie' equivalents, which may explain why evolution has favoured them and made us conscious.

Another question answered by IIT is why we are unconscious during seizures, sedation and deep sleep, but not REM sleep. Although our neurons remain alive and well during sedation and deep sleep, their interactions are weakened in a way that reduces integration and hence consciousness. During a seizure, the interactions instead get so strong that vast numbers of neurons start imitating one another, losing their ability to contribute independent information, which is another key requirement for consciousness according to IIT. This is analogous to a computer hard drive where the bits that encode information are forced to be either all zeros or all ones, resulting in the drive storing only a single bit of information. Tononi, together with Adenauer Casali, Marcello Massimini and other collaborators, validated these ideas with lab experiments in 2013. They defined a 'consciousness index' that they could measure by using an EEG to monitor the electrical activity in people's brains after magnetic stimulation, and used it to successfully predict whether they were conscious.

Awake and dreaming people had comparably high consciousness indices, whereas those anaesthetised or in deep sleep had much lower values. The index even successfully identified as conscious two patients with locked-in syndrome, who were aware and awake but prevented by paralysis from speaking or moving. This illustrates the promise of this technique for helping doctors determine whether unresponsive patients are conscious.

Despite these successes, IIT leaves many questions unanswered. If it is to extend our consciousness-detection ability to animals, computers and arbitrary physical systems, then we need to ground its principles in fundamental physics. IIT takes information measured in bits as a starting point. But when I view a brain or computer through my physicist's eyes, as myriad moving particles, then what physical properties of the system should be interpreted as logical bits of information? I interpret as a 'bit' both the position of certain electrons in my computer's RAM memory (determining whether the micro-capacitor is charged) and the position of certain sodium ions in your brain (determining whether a neuron is firing), but on the basis of what principle? Surely there should be some way of identifying consciousness from the particle motions alone, even without this information interpretation? If so, what aspects of the behaviour of particles correspond to conscious integrated information?

The problem of identifying consciousness in an arbitrary collection of moving particles is similar to the simpler problem of identifying objects in such a system. For instance, when you drink iced water, you perceive an ice cube in your glass as a separate object because its parts are more strongly connected to one another than to their environment. In other words, the ice cube is both fairly *integrated* and fairly *independent* of the liquid in the glass.

The same can be said about the ice cube's constituents, from water molecules all the way down to atoms, protons, neutrons, electrons and quarks. Zooming out, you similarly perceive the macroscopic world as a dynamic hierarchy of objects that are strongly integrated and relatively independent, all the way up to planets, solar systems and galaxies.

This grouping of particles into objects reflects how they are stuck together, which can be quantified by the amount of energy needed to pull them apart. But we can also reinterpret this in terms of information: if you know the position of one of the atoms in the piston of an engine, then this gives you information about the whereabouts of all the other atoms in the piston, because they all move together as a single object. A key difference between inanimate and conscious objects is that for the latter, too much integration is a bad thing: the piston atoms act much like neurons during a seizure, slavishly tracking one another so that very few bits of independent information exist in this system. A conscious system must thus strike a balance between too little integration (such as a liquid with atoms moving fairly independently) and too much integration (such as a solid). This suggests that consciousness is maximised near a phase transition between less- and more-ordered states; indeed, humans lose consciousness unless key physical parameters of our brain are kept within a narrow range of values.

An elegant balance between information and integration can be achieved using error-correcting codes: methods for storing bits of information that know about each other, so that all information can be recovered from a fraction of the bits. These are widely used in telecommunications, as well as in the ubiquitous QR codes from whose characteristic pattern of black and white squares your smartphone can read a web address. As error correction has proved so useful in our technology, it would be interesting to search for error-correcting codes in the brain, in case evolution has independently discovered their utility – and perhaps made us conscious as a side effect.

We know that our brains have some ability to correct errors, because you can recall the correct lyrics for a song you know from a slightly incorrect fragment of it. John Hopfield, a biophysicist renowned for his eponymous neural network model of the brain, proved that his model has precisely this error-correcting property. However, if the hundred billion neurons in our brain do form a Hopfield network, calculations show that it could only support about 37 bits of integrated information – the equivalent of a few words of text. This raises the question of why the information content of our conscious experience seems to be significantly larger than 37 bits. The plot thickens when we view our brain's moving particles as a quantum-mechanical system. As I showed in 2015, the maximum amount of integrated information then drops from 37 bits to about 0.25 bits, and making the system larger doesn't help.

This problem can be circumvented by adding another principle to the list that a physical system must obey in order to be conscious. So far I have outlined three: the *information principle* (it must have substantial information storage capacity), the *independence principle* (it must have substantial independence from the rest of the world) and the *integration principle* (it cannot consist of nearly independent parts). The aforementioned 0.25 bit problem can be bypassed if we also add the *dynamics principle* – that a conscious system must have substantial information-processing capacity, and it is this processing rather than the static information that must be integrated. For example, two separate computers or brains can't form a single consciousness.

These principles are intended as necessary but not sufficient conditions for consciousness, much like low compressibility is a necessary but not sufficient condition for being a liquid. As I explore in my book *Our Mathematical Universe*, this leads to promising prospects for grounding consciousness and IIT in fundamental physics, although much work remains and the jury is still out on whether it will succeed.

If it does succeed, this will be important not only for neuroscience and psychology, but also for fundamental physics, where many of our most glaring problems reflect our confusion about how to treat consciousness. In Einstein's theory of general relativity, we model the 'observer' as a fictitious disembodied massless entity having no effect whatsoever on that which is observed. In contrast, the textbook interpretation of quantum mechanics states that the observer does affect the observed. Yet after a century of spirited debate, there is still no consensus on how exactly to think of the quantum observer. Some recent papers have argued that the observer is the key to understanding other fundamental physics mysteries, such as why our universe appears so orderly, why time seems to have a preferred forward direction, and even why time appears to flow at all.

If we can figure out how to identify conscious observers in any physical system and calculate how they will perceive their world, then this might answer these vexing questions.

Will artificial starlight power the world?

? Even if we finally achieve the dream of controlled nuclear fusion on Earth, it will carry an environmental cost, says Jeff Hecht.

We already live in a world powered by nuclear fusion. Unfortunately the reactor is 150 million kilometres away and we haven't worked out an efficient way to tap it directly. So we burn its fossilised energy – coal, oil and gas – which is slowly boiling the planet alive, like a frog in a pan of water.

Recreating the sun on Earth would go a long way to solving that problem, but it is a biggie. Research started more than sixty years ago; the leading fusion reactor design, the tokamak, is half a century old. Tokamaks trap heavy hydrogen isotopes inside a doughnut-shaped magnetic field, heating and squeezing the plasma so that deuterium and tritium fuse to release energy. After testing a series of increasingly large tokamaks, fusion researchers agreed about ten years ago to build a huge one in France called ITER.

If everything goes to plan – which it almost certainly won't – in 2035 ITER will produce 500 megawatts of energy for a few hundred seconds. That will make it the first fusion reactor to produce more energy than it takes to operate.

Even then, two big hurdles will remain, says Mickey Wade, senior research fellow at General Atomics and former director of the US-based DIII-D fusion programme. One is developing materials that can withstand prolonged exposure to plasmas. The other is sustaining the intense magnetic fields needed to confine the plasma. Cracking all three would be an epoch-making breakthrough. Fusion would largely free us from fossil fuels, delivering clean and extremely cheap energy in almost unlimited quantities.

Or would it? Fusion power would certainly be cleaner than burning fossil fuels, but it wouldn't be carbon neutral. The reactors do not emit carbon directly, but construction, fuel production and waste management inevitably have carbon footprints of their own. Fusion also creates radioactive waste, albeit a type that decays in decades rather than hundreds or thousands of years.

Nor will fusion power be too cheap to meter. The reactors are astronomically costly to build – ITER has ballooned to more than €20 billion. Nobody will spend that sort of money if they can't recoup their investment. But once up and running, operating costs will be modest. The oceans contain enough deuterium to fuel fusion reactors for tens of thousands of years. Tritium is extremely rare in nature but can be easily made from lithium, which is also abundant.

Could the whole world run on fusion power? In principle yes, but it is unlikely in practice. Operators will want to run fusion plants as much as they can to recover their investment, so they will probably generate mostly baseload power. Peaks in demand will probably have to be met by energy storage technologies such as ultracapacitors, charged up by solar and wind. We will also have to think up new ways to power planes and other technologies that cannot run directly on grid power.

Fusion could still revert to type and remain a future technology sixty years from now. Solar and wind are unlikely to satisfy all of our needs. In that case, we may have to default to nuclear fission, with all of its downsides – accidents, long-lived waste and weapons proliferation worries. Superconductivity and geoengineering may come to our rescue. But all things told, we really need plenty of home-made sunshine by 2076.

Could replicators end material scarcity?

? When machines can make anything imaginable almost for free – the real-life *Star Trek* replicator – concepts of ownership and work will be radically transformed, says Sally Adee.

It's surprisingly hard to imagine a world without scarcity. When we think about the end of material needs, it's usually our own, says John Quiggin, an economist at the University of Queensland in Brisbane, Australia. But what about everyone's needs? 'Scarcity is the basis of our fundamental economic system,' he says. This is the capitalist paradigm whose principles are, to most of us, as non-negotiable as the laws of physics. How would the economy work if everything was free? Who would make things if no one got paid? Isn't this just communism? Trying to envision a world not organised around the market is a bit like a fish thinking about what's outside the water.

Jeremy Rifkin did it in his 2014 manifesto *The Zero Marginal Cost Society*. Capitalism, he contends, is almost done eating itself. 'It's the ultimate triumph of the market' – a final transition to a society in which automation has brought the cost of producing each additional unit of anything near to zero, and products are essentially free.

For a taster of how this looks, consider the music and publishing industries. The internet has made the production and distribution of content incredibly cheap. Though painful for some, Rifkin sees this trend as the harbinger of a new paradigm that will spread to all other industries. A critical enabler will be fabrication devices that can make almost anything on demand: think of today's 3D printers but immensely more sophisticated, like a modern computer versus a 1960s electronic calculator.

Within sixty years, these devices might have evolved into machines called molecular assemblers. The term was coined in 1977 by Eric Drexler. He imagined a nano-fabrication device capable of manipulating individual molecules with sufficient speed and precision to produce any substance you desire. Press a button, wait a while, and out come food, medicine, clothing, bicycle parts or anything at all, materialised with minimum capital or labour.

We can't know what such a world would look like – but the outlines are becoming visible. Rifkin thinks fabricators will be the engines of a sharing economy. The concept of ownership will give way to access (think Spotify and Uber for everything); purchases will give way to printing whatever you need. Within twenty years, he says, 'I don't think capitalism is going to be the exclusive arbiter of economic life. It will share the stage with its child.'

Within sixty years, capitalism might have left the building completely. In its place will be a society in which all our basic needs are met. Rifkin calls his new economic vision 'the commons', but it goes beyond the economy – it will be the new 'water' we swim in.

You will have a job, but it won't be for money. The company you work for will be a non-profit. Your 'wealth' will be measured in social capital: your reputation as a cooperative member of the species. So when you contribute to open-source code that makes a better widget, you'll enjoy a 'payment' in the form of an improved reputation. Apps that track your contribution to the commons – whether by your input at work, your frugal use of energy, or other measures of reputation – will let you cash in your karma points for luxuries – say, an antique chair that was conspicuously not built by a fabricator. Even in the commons, we'll still be human.

Will genetically engineered people conquer the world?

? Gene editing will be routinely available to improve health, says Michael Le Page, but using it to create superhuman individuals is a more distant prospect.

It is April 2021. Tarou Yamada is born in Tokyo, making headlines around the world. 'The miracle boy,' some newspapers dub him. That's because Tarou's father is unable to produce sperm because of a mutation on his Y-chromosome. He is thus, in theory, completely infertile. Yet genetic tests confirm that Tarou is his son.

To make Tarou's conception possible, a fertility clinic took stem cells from the father, corrected the Y-chromosome mutation using CRISPR genome editing and then derived sperm cells from the corrected cells. Those edited sperm cells were then used to fertilise the mother's eggs, cementing the change into all of his nuclear DNA. In other words, Tarou Yamada is the first person whose genome has been edited.

He will not be the last. While some countries have tightened regulations banning genome editing after the news of his birth (Japan has no such law at the moment), others have decided it is justified for purposes such as allowing infertile parents to have children that are biologically their own.

Soon there are dozens of genome-edited children being born every year, then hundreds, then thousands. These children are indistinguishable from typical children of the same age, because their genomes are entirely normal.

This is how the germline genome editing revolution is likely to begin. There is much talk about such editing of heritable DNA to prevent children getting genes for diseases such as cystic fibrosis from their parents, but almost all such diseases can already be prevented by screening IVF embryos before implantation.

Why should would-be parents opt for genome editing when pre-implantation genetic diagnosis, as it's called, is safer and cheaper? PGD is only good for weeding out one or two harmful mutations at a time. With genome editing, it should become possible to make dozens of changes at a time. Once germline editing starts to be used for infertility, fertility clinics are likely to offer to tweak other genes at the same time. Opponents of genetic engineering will call this a slippery slope; for proponents it is sensible, even humanitarian, progress.

We all have hundreds of harmful mutations that increase our risk of cancer, Alzheimer's, mental disorders and so on, so why not fix the worst ones if you are at it? In fact, once it can be done safely, it is arguably immoral not to.

And why stop there? There are beneficial gene variants that make people immune to HIV or less likely to become obese, for instance. Perhaps as soon as the 2030s, some countries may allow these variants to be introduced.

Such interventions would be extremely controversial. Even more so would be adding gene variants that improve personality, intelligence or other traits that we value highly. As yet we don't know how to do that – we have yet to discover any single gene variant that makes anything like as much difference to IQ as, say, having rich parents or a good education, for example.

In fact, the brain is so complex that we may never be able to predict what effect a specific mutation has. This means introducing a brain-altering mutation that does not already occur naturally would be a huge leap in the dark, one that neither parents nor regulators should ever allow.

But genome editing can definitely make individuals less prone to all kinds of diseases. And as it starts to becomes clear that genome-edited children are on average healthier than those conceived the old-fashioned way, wealthy parents will start to opt for genome editing even if there is no pressing need for them to do so.

Will this allow the elite to give their children yet another advantage, and widen the already gaping chasm between rich and poor? Quite possibly. But let's end with an optimistic prediction: by the time *New Scientist*'s 120th anniversary comes around, many countries will routinely and uncontroversially offer genome editing to any would-be parents who want it, on the basis that the cost of the treatment is far outweighed by the savings in healthcare costs over a person's lifetime.

Will the rise of robots put humans in second place?

? Great thinkers have long feared the 'technological singularity', in which the machines we designed run rings around us. Toby Walsh explains why it probably won't happen.

However you look at it, the future appears bleak. The world is under immense stress environmentally, economically and politically. It's hard to know what to fear the most. Even our own existence is no longer certain. Threats loom from many possible directions: a giant asteroid strike, global warming, a new plague, or nanomachines going rogue and turning everything into grey goo.

Another threat is artificial intelligence. In 2014, Stephen Hawking told the BBC that 'the development of full artificial intelligence could spell the end of the human race . . . It would take off on its own, and redesign itself at an ever increasing rate. Humans, who are limited by slow biological evolution, couldn't compete, and would be superseded.' The following year, he said that AI is likely to be 'either the best or worst thing ever to happen to humanity'.

Other prominent people, including Elon Musk, Bill Gates and Steve Wozniak, have made similar predictions about the risk AI poses to humanity. Nevertheless, billions of dollars continue to be funnelled into AI research. And stunning advances are being made. In a landmark match in 2016, the Go master Lee Sedol lost 4–1 to the AlphaGo computer. In many other areas, from driving taxis on the ground to winning dogfights in the air, computers are starting to take over from humans.

Hawking's fears revolve around the idea of the technological singularity. This is the point in time at which machine intelligence starts to take off, and a new more intelligent species starts to inhabit Earth. We can trace the idea of the technological singularity back to a number of different thinkers, including John von Neumann, one of the founders of computing, and the science fiction author Vernor Vinge. The idea is roughly the same age as research into AI itself. In 1958, mathematician Stanisław Ulam wrote a tribute to the recently deceased von Neumann, in which he recalled: 'One conversation centered on the ever accelerating progress of technology and changes in the mode of human life, which gives the appearance of approaching some essential singularity . . . beyond which human affairs, as we know them, could not continue'.

More recently, the idea of a technological singularity has been popularised by Ray Kurzweil, who predicts it will happen around 2045, and Nick Bostrom, who has written a bestseller on the consequences. There are several reasons to be fearful of machines overtaking us in intelligence. Humans have become the dominant species on the planet largely because we are so intelligent. Many animals are bigger, faster or stronger than us. But we used our intelligence to invent tools, agriculture and amazing technologies like steam engines, electric motors and smartphones. These have transformed our lives and allowed us to dominate the planet.

It is therefore not surprising that machines that think – and might even think better than us – threaten to usurp us. Just as elephants, dolphins and pandas depend on our goodwill for their continued existence, our fate in turn may depend on the decisions of these superior thinking machines.

The idea of an intelligence explosion, when machines recursively improve their intelligence and thus quickly exceed human intelligence, is not a particularly wild idea. The field of computing has profited considerably from many similar exponential trends. Moore's law predicted that the number of transistors on an integrated circuit would double every two years, and it has pretty much done so for decades. So it is not unreasonable to suppose AI will also experience exponential growth.

Like many of my colleagues working in AI, I predict we are just thirty or forty years away from AI achieving superhuman intelligence. But there are several strong reasons why a technological singularity is improbable.

The 'fast-thinking dog' argument

Silicon has a significant speed advantage over our brain's wetware, and this advantage doubles every two years or so, according to Moore's law. But speed alone does not bring increased intelligence. Even if I can make my dog think faster, it is still unlikely to play chess. It doesn't have the necessary mental constructs, the language and the abstractions. Steven Pinker put this argument eloquently: 'Sheer processing power is not a pixie dust that magically solves all your problems.'

Intelligence is much more than thinking faster or longer about a problem than someone else. Of course, Moore's law has helped AI. We now learn faster, and off bigger data sets. Speedier computers will certainly help us to build artificial intelligence. But, at least for humans, intelligence depends on many other things, including years of experience and training. It is not at all clear that we can short-circuit this in silicon simply by increasing the clock speed or adding more memory.

The anthropocentric argument

The singularity supposes human intelligence is some special point to pass, some sort of tipping point. Bostrom writes: 'Human-level artificial intelligence leads quickly to greater-than-human-level artificial intelligence . . . The interval during which the machines and humans are roughly matched will likely be brief. Shortly thereafter, humans will be unable to compete intellectually with artificial minds.'

If there is one thing that we should have learned from the history of science, it is that we are not as special as we would like to believe. Copernicus taught us that the universe does not revolve around Earth. Darwin showed us that we are not so different from other apes. Watson, Crick and Franklin revealed that the same DNA code of life powers us and the simplest amoeba. And artificial intelligence will no doubt teach us that human intelligence is itself nothing special. There is no reason to suppose that human intelligence is a tipping point that once passed allows for rapid increases in intelligence.

Of course, human intelligence is a special point because we are, as far as we know, unique in being able to build artefacts that amplify our intellectual abilities. We are the only creatures on the planet with sufficient intelligence to design new intelligence, and this new intelligence will not be limited by the slow process of human reproduction and evolution. But that does not bring us to the tipping point, the point of recursive self-improvement. We have no reason to suppose that human intelligence is enough to design an artificial intelligence that is sufficiently intelligent to be the starting point for a technological singularity.

Even if we have enough intelligence to design superhuman artificial intelligence, the result may not be adequate to precipitate a technological singularity. Improving intelligence is far harder than just being intelligent.

The 'diminishing returns' argument

The idea of a technological singularity supposes that improvements to intelligence will be by a relative constant multiplier, each generation getting some fraction better than the last. However, the performance of most of our AI systems has so far been that of diminishing returns. There are often lots of low-hanging fruit at the start, but we then run into difficulties when looking for improvements. This helps explain the overly optimistic claims made by many of the early AI researchers. An AI system may be able to improve itself an infinite number of times, but the extent to which its intelligence changes overall could be bounded. For instance, if each generation only improves by half the last change, then the system will never get beyond doubling its overall intelligence.

The 'limits of intelligence' argument

There are many fundamental limits within the universe. Some are physical: you cannot accelerate past the speed of light, know both position and momentum with complete accuracy, or know when a radioactive atom will decay. Any thinking machine that we build will be limited by these physical laws. Of course, if that machine is electronic or even quantum in nature, these limits are likely to be beyond the biological and chemical limits of our human brains. Nevertheless, AI may well run into some fundamental limits. Some of these may be due to the inherent uncertainty of nature. No matter how hard we think about a problem, there may be limits to the quality of our decision-making. Even a superhuman intelligence is not going to be any better than you at predicting the result of the next EuroMillions lottery.

The 'computational complexity' argument

Finally, computer science already has a well-developed theory of how difficult it is to solve different problems. There are many computational problems for which even exponential improvements are not enough to help us solve them practically. A computer cannot analyse some code and know for sure whether it will ever stop – the 'halting problem'. Alan Turing, the father of both computing and AI, famously proved that such a problem is not computable in general, no matter how fast or smart we make the computer analysing the code. Switching to other types of device such as quantum computers will help. But these will only offer exponential improvements over classical computers, which is not enough to solve problems like Turing's halting problem. There are hypothetical hypercomputers that might break through such computational barriers. However, whether such devices could exist remains controversial.

So there are many reasons why we might never witness a technological singularity. But even without an intelligence explosion, we could end up with machines that exhibit superhuman intelligence. We might just have to program much of this painfully ourselves. If this is the case, the impact of AI on our economy, and on our society, may happen less quickly than people like Hawking fear. Nevertheless, we should start planning for that impact.

Even without a technological singularity, AI is likely to have a large effect on the nature of work. Many jobs, such as taxi and truck driving, are likely to disappear in the next decade or two. This will further increase the inequalities we see in society today. And even quite limited AI is likely to have a large influence on the nature of war. Robots will industrialise warfare, lowering the barriers to war and destabilising the current world order. They will be used by terrorists and rogue nations against us. If we don't want to end up with Terminator, we had better ban robots on the battlefield soon. If we get it right, AI will help make us all healthier, wealthier and happier. If we get it wrong, AI may well be one of the worst mistakes we ever get to make.

What if the population bomb implodes?

? We could find that fears of massive overpopulation prove unfounded, and instead we live in a world where children are rare and most people are old, says Fred Pearce.

Could the population bomb be about to go off in the most unexpected way? Rather than a Malthusian meltdown, could we instead be on the verge of a demographic implosion?

To find out how and why, go to Japan, where a recent survey found that people are giving up on sex. Despite a life expectancy of 85 and rising, the number of Japanese people is falling thanks to a fertility rate of just 1.4 children per woman, and a reported epidemic of virginity. The population, it seems, is too busy (and too shy) to procreate.

It's catching. Half the world's nations have fertility rates below the replacement level of just over two children per woman. Countries across Europe and the Far East are teetering on a demographic cliff, with rates below 1.5. On recent trends, Germany and Italy could see their populations halve within the next sixty years. The world has hit peak child, claimed Hans Rosling at the Karolinska Institute in Stockholm, Sweden. Peak person cannot be far behind.

For now, the world's population continues to rise. From today's 7.4 billion people, we might reach 9 billion or so, mostly because of high fertility in Africa. The UN predicts a continuing upward trend, with population reaching around 11.2 billion in 2100. But this seems unlikely. After hitting the demographic doldrums, no country yet has seen its fertility recover. Many demographers expect a global crash to be under way by 2076.

Governments may try to halt the fall – but Singapore has been trying for a generation and still has the world's lowest fertility rate at 0.8. And once the number of fertile people starts to decline, reversing the trend will be very hard. The population boom will turn to bust.

What will this mean for the future of our species? By 2076, children will be rarities. For the first time in history most of us will be old. The brash, hormone-driven cultures that shaped the twentieth century seem doomed. Innovation could dry up.

It could trash our economies, too. Some economists say that Japan's repeated recessions since the 1990s are due to the burden of ever more oldies. Maybe Europe is going the same way. China could be next, as its population is set to peak in about 2030. Demographic determinists fear our species is on a slow downward spiral. We could go out not with a demographic bang but with an incontinent whimper.

Yet it may not be like that. A grey society will certainly be different. But perhaps, like today's ageing rock stars, we will find that being old isn't so bad. Old could be the new young. And older societies are less prone to taking up arms. A world with fewer of us would also give the planet's ecosystems a break. Malthus would be off the agenda. Instead, ecologist Edward O. Wilson's call for a century of ecological restoration could take wing. Nature, at least, would enjoy the silver lining.

The end of us? Find out on page 220 what will happen to the last humans.

How will early settlers colonise Mars?

? Overcoming all the challenges of colonising the Red Planet is a huge feat, admits Lisa Grossman, but the pioneers won't live a gilded existence.

The year is 2066. The sun rises dimly in a rust-coloured sky, lighting up the hydroponic fields. In the first permanent habitat on Mars, intrepid explorers are waking up to start another 24.5-hour day.

Elon Musk thinks it's possible. In 2016, the SpaceX founder unveiled his somewhat vague plan for sending humans to Mars in the next decade or so, and suggested that we might have a million people living full-time on the Red Planet by the 2060s. NASA's more conservative plan sees the first humans going there in the 2030s.

We'll have to get moving. Before settlers can start building a life, we would need to set up everything they need to merely survive on the surface. This means launching tonnes of life-support equipment, habitats, energy-generation systems, food, and technology for extracting breathable oxygen and drinkable water from the air.

That's a huge challenge. The shortest journey time between Earth and Mars is roughly five months, but that would only be possible around once every two years when the planets align with one another. In the most optimistic scenario, this gives us about twenty-two ideal launch opportunities to lay the groundwork for human settlement by 2060.

And as the recent failure of the ExoMars lander shows, landing on Mars is tricky: it has enough gravity to accelerate a craft's descent, but such a thin atmosphere that parachutes won't slow it down enough. The heaviest thing that ever landed on Mars, the 1-tonne Curiosity rover, used a combination of parachutes, retrorockets and a daredevil dangling device called a sky crane.

Given that we don't know how to land a mass heavier than that on the surface, planners have their work cut out. SpaceX plans to use a technique called supersonic retro-propulsion – basically reversing down with the booster rocket firing to slow the descent – and hopes to test the system in 2018. NASA has agreed to help with the project, in exchange for access to some of what we learn from it.

That's not to mention the hazards of the journey and after landing. These include high levels of radiation, the threat of solar flares, dust that covers solar panels and could rip through lungs like shards of glass, and temperatures as low as –125°C. And we don't know how to grow food there.

But let's assume we overcome all these challenges. Then what? Fans of space exploration like to point out that humans have set off from their homes in search of a new life somewhere remote and possibly dangerous many times. Getting on a boat for the New World often meant you would never see your home or family again.

What's different about Mars is that there is nothing to do there except try not to die. When European explorers struck out for the Americas, they hoped to find resources that they could sell back to their homeland, or at least a spot to establish a farm. Mars has few resources. The first settlers will be dependent on the home world for a very long time. Self-sufficiency by 2060 seems very ambitious.

One thing that settlers could usefully do, though, is science. A human could do research in an hour that takes a Mars rover months. And, of course, research on growing food would take on much more urgency.

We do have a model for such a remote yet invaluable research outpost: Antarctica. No one lives there permanently, but people make sojourns lasting a year or two to do science that is not possible anywhere else. Mars could be similar.

Another difference from past expansions into terra incognita is that Mars settlers will be in constant communication with Earth, albeit delayed a few minutes by the limits of light speed. Those of us still on Earth will almost certainly watch their lives unfold. We will see everything that goes right, and everything that goes wrong.

Whether we push farther into the solar system or retreat back to Earth will probably depend on the balance of those two things. If we figure out how to get food and air and a way of living on the Red Planet, it seems likely that we could adapt those ideas for other planets or, more probably, moons. Mars will be the first big test of whether we can become a multiplanetary species.

Where next? Turn to page 204 to find out where humans will head in millennia to come.

What will our descendants know about us?

? Bob Holmes asks what clues to the way we live today archaeologists will unearth in the millennia to come. What will endure, and what will fade away?

When humans in the far future are piecing together a picture of the primitive civilisation of 2012, archaeology will surely be the best way to go about it. After all, the best libraries, archives and museums can be undone by a single fire, amply illustrated by the fate of the library of Alexandria.

So what will archaeologists working 100,000 years from now discover about us? Only the luckiest of artefacts will avoid being crushed, scattered, recycled or decomposed. You, personally, will almost certainly leave nothing behind that survives that long. To get a sense of why, just point time's arrow the same distance in the opposite direction. Around 100,000 years ago, anatomically modern humans were just emerging from Africa to populate the world. Most of what we know about them is guesswork, because the only clues that remain are sharp stone tools and a handful of fossils.

You are especially unlikely to leave your bones behind. Fossilisation is an exceedingly rare event, especially for terrestrial animals like us – though with 7 billion people on the planet, at least a few of us will no doubt achieve lasting fame.

Luckiest – and rarest – will be the 'instant fossils'. These form when people or animals die in calcium-rich seasonal ponds and wetlands, or in caves. In both situations, bones can mineralise quickly enough for fossilisation to win the race against decomposition, says Kay Behrensmeyer, a palaeobiologist with the US National Museum of Natural History in Washington, DC. One wildebeest toe-bone in southern

Kenya soaked up calcium carbonate so quickly that it began to turn to stone within two years of death.

Future fossil hunters won't be looking for us in graveyards because bodies buried there crumble into dust within a few centuries. Instead, the richest human bonebeds are likely to be found in the debris of catastrophic events, such as volcanic ash or the fine sediments left by the recent tsunamis in Asia, Behrensmeyer says. A few bodies might be mummified in peat bogs or high deserts, but they will decay if conditions change, as is likely over a span of 100,000 years.

Those same changes will also lay waste to other important clues to our civilisation: our homes. Climate change and rising sea levels are likely to drown coastal cities such as New Orleans and Amsterdam. In these cases, waves will probably destroy the parts of buildings above ground, and basements and pilings will soon be buried by sediments. While concrete may dissolve over the millennia, archaeologists will recognise the precise rectangular patterns of sand and gravel that remain as a sign of purposeful design. 'There is nothing at all in nature like the patterns we make,' says Jan Zalasiewicz, a geologist at the University of Leicester, UK.

Nowhere will these designs be more unmistakable than in our biggest structures. A few human artefacts, such as open-pit mines, are essentially geological features already, and will last for hundreds of thousands of years as testimony to our earth-moving powers. Our largest dams, such as the Hoover dam in the US and China's Three Gorges dam, contain such an immense volume of concrete that some pieces will certainly survive that long, too, says Alexander Rose, executive director of the Long Now Foundation, based in San Francisco, California. A few structures – most notably the Onkalo nuclear waste repository in Olkiluoto, Finland – are even being engineered to survive intact for 100,000 years.

We have also been busy building another massive legacy that will be the real bumper crop for future archaeologists: our garbage. The landfill sites where most of our goods eventually end up are almost ideal places for long-term preservation. When full, modern landfills are typically sealed with an impermeable layer of clay, so that the contents quickly become devoid of oxygen, the biggest enemy of preservation. 'I think it's fair to say that these sites will remain anaerobic over geological time,' says Morton Barlaz at North Carolina State University in Raleigh. Under such conditions, even some organic materials such as natural fabrics and wood are likely to avoid decomposition – though over the millennia they will gradually transform into something resembling peat or soft coal, says landfill expert Jean Bogner of the University of Illinois at Chicago.

A few materials will be preserved just as they are. We don't make much from stone any more, but a few statues might survive, buried safely away from erosion. Ceramic plates and coffee mugs should last indefinitely, too, just as the potsherds of early human civilisations have. Some metals, such as iron, will corrode quickly, but titanium, stainless steel, gold and others will last much longer. King Tut's gold, after all, looks almost unchanged after 5,000 years. 'There's no reason to think that wouldn't be the same after 100,000 years,' says Rose. Indeed, titanium laptop cases, their insides long since corroded, may end up as one of our civilisation's most lasting artefacts. Who knows – scholars of the future may construct elaborate theories about our religious practices based on these hollow tablets and the apple-shaped figure etched into their surface.

The fact is that no matter how much we may try to preserve a legacy for future generations, we can never know which aspects of our civilisation will interest our descendants.

Today, for example, our study of early humans is informed by Darwin's theories, a perspective that was inconceivable only a century ago. Even if the objects in our museums survive, they will only tell future generations what we thought of ourselves. What they will think of us is something no one reading these words today can fathom.

6 AD 100,000: Journey to the Deep Future

Given our qualms over making predictions for the next sixty years, taking a view on the deep future might seem downright foolhardy. Yet humans have been around for 100,000 years, which is reason enough to believe we have the perspective necessary to take on the next 100,000.

Knowing how long-term forces and trends have shaped humanity and Earth, we can make intelligent predictions about what will happen next. Indeed, many groups are now attempting to extend humanity's horizons far beyond the next century, from the Long Now Foundation to those who say our presence is forging a new geological era, the Anthropocene.

In this chapter, *New Scientist* tours the coming epoch, from the language we will speak to what our descendants will make of our trash. The deep future is only just beginning.

Will we still be here?

? Asteroid strikes and supervolcano eruptions may threaten us in the next 100,000 years, but the odds are good that humanity will avoid extinction, says Michael Brooks.

What are the odds we will avoid extinction? In 2008, researchers attending the Global Catastrophic Risk Conference in Oxford, UK, took part in an informal survey of what they thought were the risks to humanity. They gave humans only a 19 per cent chance of surviving until 2100. Yet when you look more closely, such extreme pessimism is unfounded. Not only will we survive to 2100, it's overwhelmingly likely that we'll survive for at least the next 100,000 years.

Take calculations by J. Richard Gott, an astrophysicist at Princeton University. Based on 200,000 years of human existence, he estimates we will likely last anywhere from another 5,100 to 7.8 million years. Fossil evidence is similarly reassuring. Records in the rocks suggest that the average species survival time for mammals is about a million years, though some species survive ten times as long. It seems there is plenty of time left on our clock. Plus, if you'll excuse the blowing of our own trumpet, we are the cleverest of the mammals.

Mind you, this could be seen as a problem. Probably the greatest threat to an advanced civilisation is technology that runs out of control; nuclear weapons, bioengineering and nanotechnology have all been cited as bogeymen. But disaster expert Jared Diamond, a geographer at the University of California, Los Angeles, points out that we no longer live in isolated civilisations. Humanity is now a global network of civilisations, with unprecedented access to a diverse, hard-won pool of knowledge already being harnessed for everyone's protection.

We are also unlikely to be extinguished by a killer virus pandemic. The worst pandemics occur when a new strain of flu virus spreads across the globe. In this scenario people have no immunity, leaving large populations exposed. Four such events have occurred in the last 100 years – the worst, the 1918 flu pandemic, killed less than 6 per cent of the world's population. More will come, but disease-led extinctions of an entire species only occur when the population is confined to a small area, such as an island. A severe outbreak will kill many millions but there is no compelling reason to think any future virus mutations will trigger our total demise.

More scary is the prospect of a supervolcano eruption. Every 50,000 years or so, a supervolcano somewhere erupts and ejects more than 1,000 cubic kilometres of ash. Such events have been linked with crashes in human population. Around 74,000 years ago, Toba erupted in Sumatra. Anthropologists have suggested that the event may have reduced the human population of Earth to just a few thousand. But as Bill McGuire, director of the Benfield Hazard Research Centre at University College London, points out, there were many fewer humans then and they were largely confined to the tropics, a geographical concentration that made the eruption's impact much more severe than would be the case with today's widely distributed population. 'Wiping out seven billion people today would be far more difficult,' he says.

Judging by their historical frequency, it is estimated that the chance of a super-eruption in the next 100,000 years is between 10 and 20 per cent. With colossal clouds of ash plunging the surface of Earth into darkness for five or six years, global harvests would be badly hit for long enough to cause loss of life on an unprecedented scale. 'The likely death toll would be in the billions,' McGuire says. But it would have to happen twice in that timescale for a realistic chance of human extinction. That's not impossible, just statistically extremely unlikely.'

The biggest extinction threats of all come from space. Solar flares, asteroid strikes and bursts of gamma rays from supernova explosions or collapsing stars are what we really need to get through. 'Every three hundred million years we would expect a gamma-ray burst or a severe supernova explosion that wipes out most of the ozone layer,' says Brian Thomas, an expert on intergalactic hazards based at Washburn University in Topeka, Kansas. The result would be a massive increase in harmful radiation at Earth's surface and an increased incidence of life-threatening cancers during the decades it would take for the ozone layer to recover. It's impossible to know when such an event might occur.

Yet these things are so rare that the chance of an extinction event in the next 100,000 years is effectively zero. The same can be said for the threat of a solar flare so powerful that it knocks out all critical infrastructure, because it would take flares 1,000 times more powerful than the biggest ever seen. 'Can our sun, in its present state, produce such a flare very occasionally? We don't know,' says Mike Hapgood, a solar physicist based at the Rutherford Appleton Laboratory in Oxford, UK, and project manager for the European Space Agency's Space Weather Programme. But it remains an unlikely disaster scenario. Which leaves the poster child of disaster movies: the asteroid strike.

This one will take some luck to avoid. Space is full of rocky debris that acts as an occasional threat to Earth. It is widely believed that the impact of a 15-kilometre-wide asteroid wiped out the dinosaurs 65 million years ago. In any 100,000-year period we can reasonably expect an impact from a 400-metre asteroid that will cause damage equivalent to 10,000 megatonnes of TNT. 'Not enough to do in the whole civilisation, but certainly destroy an entire small country like France,' says former astronaut Thomas Jones, who co-chairs NASA's Task Force on Planetary Defence.

Some might argue that without France there is little hope

for civilisation anyway, but in reality there is only a one-in-five chance of total wipeout. 'Global effects come from an impact roughly every 500,000 years, so the odds are about 20 per cent for a catastrophic, civilisation-threatening impact within 100,000 years,' Jones says. We should probably work on some anti-asteroid measures, but really, humans concerned about the longevity of our species can relax: the view from here is fine.

What will future humans be like?

? We're not so different from humans who roamed Earth 30,000 years ago, says Graham Lawton. Will genetic engineering transform us in the long run?

There's a famous thought experiment about kidnapping a Cro-Magnon man, bathing and shaving him, dressing him in a suit and putting him on the New York subway. Would anybody bat an eyelid?

Probably not. Though Cro-Magnons lived about 30,000 years ago, they were to all intents and purposes modern humans. Physically they were perhaps a little more robust, but behaviourally they were indistinguishable from us, give or take the effects of thousands of years of technological progress on our lives.

We have undoubtedly come a long way since then. A Cro-Magnon in twenty-first-century New York would recognise almost nothing except for other human beings. But his modern human brain would eventually adjust to the startling new surroundings, much as the Tierra del Fuego native who became known as 'Jeremy Button' took to Victorian London after he was brought there in 1830 by Robert FitzRoy, captain of the *Beagle*.

Now turn that thought experiment on its head and project it into the deep future. What if somebody alive today could be transported to the equivalent of New York 30,000 years – or even 100,000 years – from now? Even if suitably attired, would they fit in?

Impossible to say, of course. Just because we've had more than a thousand generations of biological stasis does not mean we can expect thousands more. If you believe some futurists, we will eventually become cyborgs with prostheses in our brains and nanobots racing around our bloodstream.

Extreme as these technological enhancements may sound, they won't produce changes to our bodies and minds that will be heritable and so alter our fundamental biology. Each generation will have to choose whether or not to become cyborgs, just as people can opt for laser eye surgery today. For our descendants to be radically different from us, we would have to engineer our own genome or wait for an event that has happened only rarely in our evolutionary line.

One hypothesis to explain the sudden rise in behaviourally modern humans 30,000 to 40,000 years ago is the random appearance of a beneficial genetic mutation, perhaps involved in language. So beneficial, in fact, that the mutation swept through the population. Humans without it would have been unable to compete with their more fortunate fellows, and their less fit genomes would have been consigned to the scrapheap of evolution.

The 'great leap forward' mutation, if it ever existed, will probably never be identified as it has completely replaced the version of the gene that preceded it. But we can see signs of similar sweeps that are not yet complete. For example, a mutation in a gene called microcephalin arose around 14,000 years ago and is now carried by 70 per cent of people. It appears to be involved in brain development, though it is not clear what trait it is being selected for since there is no discernible difference between people who carry it and those who don't.

So it is possible that our descendants could evolve into something similar to *Homo sapiens* today. But radical change seems a long shot.

Of course, we could eventually decide to take evolution into our own hands. In principle, we could engineer ourselves into obsolescence by creating a new breed of human that would outcompete ourselves. The most plausible technology for starting down this road is to genetically engineer sperm or eggs, or early embryos, in order to install changes in their genomes that will be passed down the generations. This is just about possible with today's technology, and has been put forward as a way of stamping out genetic diseases such as cystic fibrosis.

Would we go so far as to put desirable traits in rather than just take bad ones out? Even if it were technically possible to do this, it is doubtful that we would collectively agree such changes on a scale that would alter the course of our evolution – unless, of course, engineered humans were so superior that they obliterated the competition.

These possibilities cannot be ruled out. Surely the most likely option is that our time traveller will find himself among friends, a species of human fundamentally the same as us but with cooler technology. Deep down, they will still be human.

How will our language evolve?

? Given the rapid change in language in just a few millennia, what will it be like tens of thousands of years from now, wonders David Robson.

Should your descendants uncover this page, yellowed and curling, thousands of years from now, many of these words will be incomprehensible – even if they call themselves speakers of English. After all, we struggle to decipher old English texts like *Beowulf*. You might be able to understand the hero's declaration that 'Béowulf is mín nama', but a millennium of language evolution has washed away the meaning from 'grimma gaést Grendel' – the 'ghastly demon Grendel'.

If our language has transformed almost beyond recognition in just 1,000 years, how might it sound in tens of thousands of years? Languages are largely shaped by the unpredictable whims of their speakers, but by examining the forces facing our language, we can speculate about how our descendants might speak.

The most obvious question is whether they will be using English at all. Although English is the world's lingua franca, its popularity largely hinges on the present economic importance of anglophone countries. Should another country come to dominate world trade, our descendants may all be learning its language. If so, it's likely that they would begin to incorporate some of its terms into their own language – in the same way that Italians say that they will listen to a 'podcast' on their 'tablet' at the 'weekend'. But very popular languages tend to be resilient to invasion, so there's no reason to think that English will disappear entirely.

It's more likely that it will splinter and fragment. We can already see new dialects forming in many of the UK's former colonial territories, such as Singapore and Jamaica. Thanks to immigration, the internet and mass media, words from such dialects often feed back through the English-speaking world – as can be seen in Jamaican variations that are now sweeping through London slang, such as the use of 'buff' to mean attractive, and 'batty' to mean a person's bottom. Given enough time, these dialects might diverge entirely. If so, English may end up like Latin – dead, but survived by numerous offspring.

Do such grand transformations make it impossible to predict anything specific about future English? Certainly, the language is changing quickly enough as it is; the *Oxford English Dictionary* adds between 2,000 and 2,500 words each year, says its senior assistant editor, Denny Hilton. But there may be thousands of new words that fail to catch the attention of the OED's lexicographers. When Erez Lieberman Aiden and Jean-Baptiste Michel at Harvard University studied Google's corpus of digitised books from the last century, they found around 8,500 new words entering the language every year. Many of these are rarely used – words like postcoarctation, reptating and subsectoral.

By looking at English's journey since *Beowulf*, we can at least identify trends that might continue. Its future grammar might lack some of the nuances that rule the sentences on this page, for instance. We've already lost many of the rules that governed the language of *Beowulf* – English nouns no longer have different genders, for instance.

Today, this ongoing simplification can be seen in the way we use the past tense. There are lots of irregular verbs whose past tenses do not have the more typical '-ed' ending – we say 'left' rather than 'leaved', for example. But time is slowly taming these irregular verbs, and the effect depends on how common these verbs are. By studying English texts from the last 1,000 years, Lieberman Aiden and Michel noticed that the less a verb is used, the more likely it is to become regular. 'If a word is rare, we don't always remember if it is irregular,' says Lieberman Aiden – so we assume it follows the pattern of more familiar verbs.

'To wed', which is now used only in very specific contexts,

is already in the throes of change. People are beginning to say they are 'newly wedded' rather than 'newly wed', for example. Others are more stubborn. Having found the way a word's popularity can influence its chances of linguistic change, Lieberman Aiden and Michel started to predict the future lifespan of certain irregular verbs. For instance, given its relative rarity, there is a 50 per cent chance that 'slunk' will become 'slinked' within 300 years.

'To be' or 'to have', which are used in around one in ten sentences, have 'half-lives' of nearly 40,000 years. The researchers speculate that irregular plurals will follow a similar trend – 'men' could become 'mans', for example – though they haven't tested the idea yet.

In a similar way, we can predict which words will be ousted by new coinages or terms imported from another language. By examining linguistic evolution across the Indo-European languages, Mark Pagel at the University of Reading, UK, has found that this too depends on a word's frequency – the more common it is, the longer it lingers. That's partly because we are less likely to use the wrong term if we hear the right term often enough.

In his book *Wired for Culture*, Pagel also argues that words have evolved to suit their purpose – if they are common and represent important concepts, they will be short and easy to say. Such words are 'highly fit', he says, using a Darwinian analogy. 'It's difficult for a new word to dislodge them.'

This can be seen in Beowulf's declaration. 'Nama' clearly lingers as 'name', a very common word then and now. Numbers, question-words and other simple nouns have similar staying power.

So, if your descendants do speak a form of English and happen to be reading this page, there's a chance they may find some meaning in simple sentences like 'What is your name?' or 'I drink water'. There's a slim chance they might even comprehend 'Hello from the year 2017'.

Where will we live?

? Plate tectonics, volcanoes and rising seas will reshape our world. Michael Le Page and Jeff Hecht ask what it will look like, and where we will live.

Fishing boats in the North Sea bring up some strange things in their nets, from the bones of mammoths to ancient stone tools and weapons. Here and in many other places around the world, we are discovering the remains of human settlements on what is now the seabed. As the world changed after the last ice age, many of our ancestors were forced to abandon their homes. And over the next 1,000 years, let alone 100,000, the world is going to change dramatically again, forcing billions of people to find a new place to live.

Some places would battle to survive even if sea levels remained constant. The ancient Egyptian city of Herakleion disappeared beneath the Mediterranean Sea 2,000 years ago as the soft sands of the delta it was built on subsided, and the same is happening to modern cities such as New Orleans and Shanghai. In Miami and elsewhere, seas and rivers are eroding the land that cities are built on.

With a stable climate, it might be possible to save cities like these. But as the world continues to warm, rising sea levels are going to drown many of our coastal cities, along with much farmland. The changing climate will also affect people living well above sea level, making some areas uninhabitable but creating new opportunities elsewhere.

We don't know exactly how much hotter the world will

become. But let's suppose events follow the Intergovernmental Panel on Climate Change's 'business as usual' scenario, with greenhouse emissions continuing to grow until 2100 and then declining rapidly. Suppose, too, that we do not attempt any kind of geoengineering.

The most likely result is that the average global temperature will rise nearly 4°C above the pre-industrial level around the year 2100, peaking at 5°C some time in the twenty-third century (though it might well get a lot hotter than this). It will stay hot, too, as it will take 3,000 years or so for the planet to cool just 1°C.

That might mean that the Greenland ice sheet will be almost gone in 1,000 years, with the West Antarctic ice sheet following it into the sea, raising its level by well over ten metres. That's bad news given that coastal regions are home to much of the world's population, including many rapidly growing megacities. As the sea level rises, billions of people will be displaced.

At least this will likely be a gradual process, though there may be occasional catastrophes when storm surges overcome flood defences. Large areas of Florida, the East and Gulf coasts of the US, the Netherlands and the UK will eventually be inundated. Some island nations will simply cease to exist and many of the world's greatest cities, including London, New York and Tokyo, will be partly or entirely lost beneath the waves.

And as the great ice sheet of East Antarctica slowly melts, the sea will rise even higher. For each 1°C increase in temperature, sea level could eventually rise by 5 to 20 metres. So in 5,000 years' time, the sea could be well over forty metres higher than today.

Even those living well above sea level may be forced to move. Some regions, including parts of the southern US, may become too dry to support farming or large cities. In other areas, flooding may drive people out.

Any further warming will cause catastrophic problems. A 7°C global rise will make some tropical regions so hot and humid that humans will not be able to survive without air conditioning. If the world warms by 11°C, much of the eastern US, China, Australia and South America, and the entire Indian subcontinent, will become uninhabitable.

Yet the future will open up alternative places to live. In the far north, what is now barren tundra and taiga could become fertile farmland. New land will also appear as the ice sheets melt. A rush to exploit the resources in newly exposed bedrock in Antarctica, for instance, could encourage settlement in its coastal regions. If it stays hot enough for long enough, Antarctica will once again be a lush green continent covered in forests. Elsewhere, pockets of fresh land will rise out of the ocean in the space of hundreds of thousands of years, perhaps ripe for human settlement.

At some point our descendants could take control of the global climate. But it will take thousands of years to restore the ice sheets and get sea levels back down. By the time we are in a position to do so, some people may like life just as it is. The proud citizens of the Republic of Antarctica will fight any measure that would lead to their farms and cities being crushed by ice.

Throughout history, explorers have planted their flags on virgin lands. Today, there's almost nowhere left on Earth where we haven't set foot – but that won't always be the case. Plate tectonics and volcanism are continually creating new land. For example, future settlers are likely to find Hawaii has an extra island. For more than eighty million years, a 'hot spot' of rising magma from deep within Earth has punched through the floor of the Pacific Ocean to build a series of islands on the crust moving over it. This means Hawaii's Big Island will soon get a baby brother off its south coast, formed by a submerged volcano called Lo'ihi. It is growing fast and should emerge within 100,000 years, depending on sea-level rise. Geologists expect that its peak will eventually tower above all others in the Hawaiian chain.

In the much longer term, Europe and Africa could also get swathes of new territory. That's because Africa is moving north-east by about 2.5 centimetres a year, gaining about a centimetre a year on Europe, which is moving in the same direction. In principle, this crunching could shut the Strait of Gibraltar within the next few million years. Without the inflow of Atlantic water, the Mediterranean Sea would eventually evaporate. Countries in southern Europe and on the north African coast would effectively expand across the newly exposed seabed until they join up.

If our descendants are still around millions of years from now, they may have to figure out how to divvy up whole new parts of the world.

Will there be any nature left?

? We are causing a mass extinction event. What species do we stand to lose in the coming millennia, asks Michael Marshall, and what new creatures will emerge?

On the face of it, the future of the natural world looks grim. Humans are causing a mass extinction that will be among the worst in Earth's history. Wilderness is being razed and we are filling the air, water and land with pollution. The bottom line is that, barring a radical shift in human behaviour,

our distant descendants will live in a world severely depleted of nature's wonders.

Biodiversity, in particular, will be hit hard. Assessments of the state of affairs make consistently depressing reading. Almost a fifth of vertebrates are classed as threatened, meaning there is a significant chance that those species will die out within fifty years.

The main cause is habitat destruction, but human-made climate change will be increasingly important. One much-discussed model estimates that between 15 and 37 per cent of species will be 'committed to extinction' by 2050 as a result of warming.

'It will be a new world,' says Kate Jones at the Institute of Zoology in London, UK. The ecosystem will become much simpler, dominated by a small number of widespread, populous species. Among animals that are 'incompatible' with humans – we may like hunting them or colonising their habitat, for example – few will survive. 'I don't have much hope for blue macaws, pandas, rhinos or tigers,' Jones says.

Ultimately, though, life will recover: it always has. The mass extinctions of the past offer hints as to how the ecosystem will eventually bounce back, says Mike Benton at the University of Bristol, UK. The two that we know most about are the end-Permian extinction 252 million years ago, which wiped out 80 per cent of species, and the less severe end-Cretaceous extinction 65 million years ago, which famously took out the dinosaurs. The Permian extinction is more relevant because it was caused by massive global warming, but Benton cautions that the world was very different then, so today's mass extinction will not play out in quite the same way.

Recoveries usually have two stages. If ours pans out in the same way, the first 2 to 3 million years will be dominated by fast-reproducing, short-lived 'disaster taxa'. These will rapidly give rise to new species and bring the world's species count back up.

But a lot of things will still be missing. Ecosystems will be simple, with similar species doing similar things. Herbivores will be less diverse, and top predators may be absent altogether in many places. That's where longer-lived, slower-evolving species come in to restore the full complexity of the ecosystem. But this can take up to ten million years, much longer than even the most optimistic projections of the human future.

It doesn't have to be like that. We can take action now to get the recovery going, although we don't know how much we can accelerate it. Conservation biologists are increasingly thinking the unthinkable, such as relocating species to places where they can thrive while abandoning them to their fate in their native ranges. That may seem unnatural, but given that human influence has already touched almost every ecosystem on Earth, is 'natural' even a useful concept any more?

Even more radically, we might be better off encouraging the formation of new species and ecosystems rather than struggling to save existing species that have no long-term future, such as pandas. 'There's no way I'd want to get rid of them,' says Jones, 'but things do change and adapt and die.'

Benton says the most important thing is to rebuild biodiversity hot spots such as rainforests and coral reefs. That needn't be a gargantuan task. A recent analysis suggests that damaged wetlands can be restored within two human generations.

Beyond that it may be possible to start 'evolutionary engineering'. For instance, we could divide a species into two separate habitats and leave them to evolve separately, or introduce 'founder' species into newly rebuilt ecosystems.

Nature may solve the problem for us by providing founder species from an unexpected source. Animals such as pigeons, rats and foxes are already flourishing alongside humans and may well give rise to new species, becoming the founders of the new ecosystem. If you are disturbed by the prospect of a world colonised by armies of rapidly evolving rats and pigeons, look away now.

Where will we explore?

? As we spread into the cosmos, the route we take will be shaped by old and familiar human desires – and perhaps even a dash of religious fervour, says Anne-Marie Corley.

It is an inescapable fact. The destinations we can visit in outer space will always be limited by the technical challenges of travelling the unimaginable distances involved, especially within a human lifespan. Still, that will not be the only factor shaping where our descendants go. The route that they take into the cosmos will be equally driven by age-old human motivations – and perhaps even a dash of religious fervour.

First, the bad news. In 2011, a group of scientists, engineers and futurists assembled in Orlando, Florida, to plot humanity's next era of exploration. The name of the plan was the 100 Year Starship Study. The idea was to begin to work out, over the next century, how to get humans to the nearest stars. You can't fault the idea for ambition, but many of them soon realised that developing the necessary technology was daunting, if not fanciful.

Neal Pellis of the Universities Space Research Association

based in Columbia, Maryland, summed up just how far our fastest spacecraft are from achieving interstellar travel. 'The nearest star is Alpha Centauri,' he told the 100 Year Starship meeting's participants. 'At 25,000 miles per hour, it would take 115,000 years to get there. So this is not a plan.'

Even if we figure out how to travel at the speeds required to arrive at a star in a human lifetime, the energy required to get there is far beyond our means for the foreseeable future. Marc Millis of the Tau Zero Foundation, a space-travel think tank based in Fairview Park, Ohio, says only a tiny proportion of today's global energy output goes towards space flight. If this state of affairs persists while energy production continues to grow at the rate of recent decades, then interstellar missions are at least two to five centuries away, he calculates.

For the next few centuries, then, if not thousands of years hence, humanity will be largely confined to the solar system. Even reaching destinations closer to home will remain slow going until we find better propulsion systems than chemical rockets, which are like Columbus's ships in terms of speed and technology, says NASA planetary scientist Chris McKay.

Assuming we achieve the speed boost we need, what routes might we take farther into space, and what will drive exploration? Scientists will no doubt continue to send uncrewed probes all over the solar system, but if history is any guide, human exploration and settlement of space will not be driven by scientific curiosity alone.

Roger Launius, NASA's former chief historian, now senior curator at the Smithsonian National Air and Space Museum in Washington, DC, says that whenever people have ventured into unexplored corners of Earth, their motivation has tended to be 'God, gold or glory' – in other words, a drive to convert indigenous peoples or escape religious persecution, or to extract wealth or earn fame.

Much of human space exploration to date has arguably been motivated by glory. National pride was behind the first crewed space missions and fuelled the colossal investment required to put people on the moon. Political will of the same order will be needed to realise the first Mars walk or human visit to an asteroid.

Farther down the track, nations or companies may want to be the first to send astronauts to rocky worlds like Saturn's moon Titan, which sports polar lakes of liquid methane. Another tempting expedition would be to Jupiter's moon Europa – especially if the liquid ocean under its surface ice turns out to be home to extreme life forms.

What about God? Could religious motivation play a role in space travel? Future solar-system explorers will have no local aliens to convert, but religion could conceivably be a reason to flee Earth. In the seventeenth century, for example, English Puritans risked their lives to settle in America for the sake of practising their beliefs. If the private space-flight industry provides the means, it's not impossible that a religious group might be among the first to populate the moon or a Mars base.

Nevertheless, the dominant drivers of exploration in our history have been economic ones, Launius says. For a space economy, mining asteroids has been proposed, as has space tourism, but neither's time has come yet. 'We have yet to find an economic motive to undertake space activities that would involve humans,' Launius says. For example, it's impossible to predict what mineral resources will be important to us mere decades from now. By the time it becomes viable to mine, say, platinum from asteroids, humanity's demand for that metal may have faded.

Another lesson of history is that exploration has not always been sustained. Instead, it often happens in fits and starts. Consider how the Vikings ventured into North America a thousand years ago, yet permanent European settlement did not follow for another four centuries. Chinese exploration also went on for centuries but ceased by 1500 or thereabouts. 'There's nothing inevitable about space travel,' says John Logsdon, a space-policy researcher at George Washington University in Washington, DC. He suggests that subsequent generations may take a break from exploring deep space or even venturing beyond Earth.

Indeed, our descendants may well have to come to terms with never having the means or lifespan to reach other stars. For them, the stars will remain tantalising twinkles of light, forever beyond reach. Then again, there will always be people, like the delegates to the 100 Year Starship meeting, who will work to keep the dream alive.

Will we run out of resources?

? Depletion of useful materials has often prompted predictions of doom, says Richard Webb, but we need not worry in the long term.

In 1924, a young mining engineer named Ira Joralemon made an impassioned address to the Commonwealth Club of California. 'The age of electricity and of copper will be short,' he said. 'At the intense rate of production that must come, the copper supply of the world will last hardly a score of years . . . Our civilisation based on electrical power will dwindle and die.'

Copper – and civilisation – are still here. Yet almost a century on from Joralemon's warning, similar wake-up calls can still be heard. The price of copper has surged to a series of all-time highs on the back of increased demand from China.

'Peak copper' is upon us, say some; reserves will run out within a couple of decades, say others.

Such prophecies of doom overlook something important. For most of our history, the way technology has developed has been determined by the materials available: think Stone Age, Bronze Age, Iron Age. But while we might label our era the Silicon Age – or perhaps, more pertinently, the Hydrocarbon Age – we are not one-trick ponies any more. These days the rapid pace of technological development is more likely to change the materials we rely on.

The *Engineering and Mining Journal-Press* drove home the point in a prescient editorial response to Joralemon's warning. 'We can hardly believe that all our electricity will go back to the clouds where Franklin found it, just because copper is scarce,' it said. 'Maybe copper won't be required at all for transmission purposes; we may just use the ether.' And indeed we do, for long-distance communications that once required large quantities of wire. We also take full advantage of optical fibres, a technology whose widespread use was hardly imagined back in the 1920s.

That makes second-guessing tomorrow's materials landscape foolhardy over timescales of mere decades, let alone millennia. 'Within fifty or sixty years we will have made so much progress that it's almost like hitting a big brick wall making any predictions beyond that,' says Ian Pearson of Futurizon, a consultancy specialising in future technologies.

The rare earth metals are a case in point. Shortages of these elements, whose applications range from touchscreens to batteries and energy-efficient light bulbs, are widely predicted within the next decade or so. Much beyond that, though, and it seems implausible to argue that we won't have innovated our way around supply bottlenecks.

'It's fashionable to talk about a shortage of neodymium for magnets in wind turbines, for example,' says Pearson, 'but the fundamental problem is not neodymium. It is how we extract energy efficiently from wind.' No doubt there are as yet undreamed-of ways to do that without building turbines. In the longer term, other innovations may render the whole idea of wind energy passé.

Whatever problems we do face in the deep future, Pearson reckons a shortage of materials is unlikely to be one of them. 'Regardless of what humanity is like in five hundred or a thousand years' time, we will probably still be filling only ten, maybe fifteen metres of air above ground with stuff,' he says. 'But there's six thousand kilometres of ground with stuff in it beneath us.' It's also plausible that it will become technologically and economically viable to mine nearby asteroids for elements we may be running short of.

To ensure the continued survival of our species, it makes sense that we should husband the resources of Earth and its environs, rather than plunder them. Technology could make that easier. Whenever we use stuff, we hardly ever export the constituent atoms and molecules beyond the Earth system; we merely rearrange them chemically, for example converting carbon locked in fossil fuels into carbon dioxide. At present, we are not particularly good at converting our waste products into something useful. But given a few more decades, things could look very different thanks to new methods for nanoscale material manipulation, as well as genetically engineered bacteria that would eat up waste and burp it out in other forms.

By then things will probably be out of our hands anyway, says Ray Hammond, an independent futurology consultant. At some point, we will create computers far more capable than ourselves. 'What these machines may be able to suggest to us in the way of resource management or in the construction of synthetic resources is wholly unknowable,' he says.

That suggests we should be worrying about other existential threats in the deep future. 'The idea that "things will run out" is to think about the future using today's concepts,' Hammond says.

7 Journey's End

And so we approach our final destination. No matter which strand of the multiverse you choose to live in, or in which direction your future path takes you, one thing is certain – sooner or later, you're going to run out of road to travel. In this chapter, we examine the shape of these destinations: what lies in wait for us at the end, whether it's the collapse of civilisation or the disintegration of reality itself. Please ensure sure your tray table is stowed and your seat is in the upright position. We'll see you at the end of the universe.

Are all societies doomed to collapse?

? Every empire falls; what hope is there that ours will be the one to buck the trend? Debora MacKenzie sifts through the ashes of former civilisations to find out what went wrong – and whether we could survive similar perils.

Rome, the Maya, Bronze Age Greece: every complex society in history has collapsed. Will our industrial civilisation be any different? Probably not. It all comes down to complexity and energy. Societies inevitably grow more complex as they chase prosperity and find solutions to the problems thrown up by success, and that comes at a cost: energy. Civilisations collapse, the thinking goes, when they can no longer generate enough juice to maintain existing complexity and solve new problems.

We got to where we are today because the industrial revolution exploited readily available high-quality anthracite coal. We then used that energy to tap progressively harder-to-access energy sources, driving our complexity to unprecedented heights. But unless we find a bounteous new source, we will one day overshoot what we can afford. Then complexity quickly unravels: political and economic institutions falter, production and trade diminish, global supply chains break. Technologies become impossible. States fragment. Lots of people die.

But there is hope. Except for small, isolated societies in which everyone died, no historical collapse has wiped the slate clean. All retained enough of their technologies and institutions to start afresh, and eventually do better. So could our descendants take what remains and build a new civilisation?

The problem is that this time there might be nothing left. 'Rome didn't have nuclear weapons,' says Ian Morris at Stanford University in California. Collapsing societies undergo dramatic shifts in power and wealth, which are always accompanied by violence, he says. 'This could be the final collapse.'

Globalisation could also make our meltdown different. When past societies fell, there were others left to carry on, says Thomas Homer-Dixon at the University of Waterloo in Canada. 'If our one global civilisation collapses there won't be outside resources, capital and knowledge to reboot things.'

For Ugo Bardi at the University of Florence in Italy, the chances of rebuilding depend on whether we can keep the electrical grid running. This isn't just to keep the lights on, but to produce the materials required for industrial civilisation – steel for machinery, potash for fertiliser, silicon for semiconductors and so on. With easily accessible fossil-fuel energy sources long exhausted, Bardi calculates that after a collapse we wouldn't be able to recover enough energy to mine or smelt the materials we rely on unless we retain a working grid.

That means we can future-proof our energy supplies, but only if we act now. Generating fossil-fuel or nuclear energy requires substantial energy up front – if that system collapses we won't have what it takes to crank it up again. Sun and wind, however, are free; we need only maintain the devices that capture them.

Bardi calculates that if half our electricity came from renewables, the grid could generate enough energy to maintain us and, crucially, itself, through crises that would completely collapse our present system. But we would need to build it while we have the silicon and civil order, and that would require investment in renewables to be fifty times its current level.

If not, says Bardi, 'we don't have enough anthracite to reinvent electricity or launch the industrial revolution again. So it will be agriculture: simple tools and dark nights.' Then again, climate instability might hinder farming, leaving hunting and gathering.

To do any better than that we will need to keep our key institutions, Homer-Dixon thinks, but that could be impossible amid severe climate change and conflict. When things settle down, all our records could be gone: even hard drives decay in a century or two.

And in case you think we might be better off forgetting the knowledge that led to our civilisation's fall, think again: the more primitive the society, the more violent people were. Collapse will be no return to Eden. Time to start installing those solar cells.

What will become of the last humans?

? There's a chance that small groups of humans could survive an Earth-shattering apocalypse. But fragmented and isolated, they would eventually become something else, writes Christopher Kemp.

We humans are a successful bunch: no other species has guided its own fate or shaped its environment as completely as we have. In doing so, we have sidestepped many of the selection pressures that drove our evolution. But *Homo sapiens* has evolved more quickly than ever in the past few thousand years and will continue to do so. So what will be the fate of our species?

Predicting our future evolution is a tricky business: it's hard to know what genetic novelties might arise, and which of them might take hold. Even so, scientists have started to weigh the possibilities by studying trends in health and reproduction.

A team led by Stephen Stearns at Yale University, for example, has found that in the sixty years since 1948, relatively short and heavy women in the town of Framingham, Massachusetts, tended to have more children than women with the opposite traits. They also found that these physical characteristics are being passed on to their daughters, suggesting natural selection is alive and well in humans. It is difficult to say what is selecting for these traits, but it looks as if we can expect women in Western countries, on average, to become slightly shorter and stouter.

More radically, we might begin to direct our own evolution. In one sense we already do: by shaping our environment and culture, we inadvertently drive heritable changes in our genes. But if sophisticated gene-editing technologies permit whole genome engineering in sperm and eggs, we would have more control than ever – we would get to choose which traits we pass on to the next generation.

It's possible we will cut the whole enterprise short by bringing about our own extinction through nuclear annihilation, runaway climate change or some other cataclysm. In most apocalyptic scenarios, however, at least a few *H. sapiens* would survive, perhaps forced to retreat into remote refuges. From a deeply interconnected species of 7 billion individuals, we would become fragmented across various ecological settings, each population beset by local environmental pressures.

These are the conditions that favour the formation of new and distinct species, says Ian Tattersall, a palaeoanthropologist at the American Museum of Natural History in New York City. If those populations are small enough, over deep time the random mutations that prove advantageous might be incorporated into the genomes of surviving *H. sapiens* – and as the genetic novelties accumulate, populations might begin to diverge.

Eventually, for instance, humans in the northern reaches of Arctic Canada might adapt to the challenges of their environment to become something new. Meanwhile, in Australia, people might adapt in profoundly different ways. Eventually, members of one group would no longer be able to mate with those of the other – one of the key signifiers of a distinct species emerging.

In the event that those new hominin species ever came into contact again, it would be war. 'We would have a situation pretty much like what we already had at the end of the last ice age,' says Tattersall. 'Modern humans spread all over the world, encountered other hominid populations and eliminated them.' So history could repeat: like those we outcompeted in the past, our all-conquering species might itself be driven to extinction.

Making a new hominin would be a slow process. It would take hundreds of thousands if not millions of years, says David Pilbeam, who studies human evolution at Harvard University – and that makes this sort of speciation unlikely. Regardless of the apocalyptic scenario, unless humans lost the urge to explore, isolated populations would encounter each other and breed before speciation could occur.

Ultimately, *H. sapiens* might have to colonise other planets to provide the long-term isolation required for speciation. So if there is ever to be a new form of human, it will be shaped by the alien environment of a strange new world in outer space.

What if everything died out tomorrow?

? Earlier in the book Bob Holmes killed every human on the planet. This time, he wipes out all life itself. Join him to see what happens next.

The end could come with a bang – a nearby supernova that bathes Earth in deadly gamma rays. Or it might come with a whimper – a supervirus that somehow proves lethal to every living cell on the planet. Neither is remotely likely, but nor are they impossible. Yet thinking about them raises an intriguing question: what would happen to Earth if every living thing were to die tomorrow?

More than you might think. Life is far more than a trivial infestation atop the physical structure of our planet. Living organisms play a major role in a wide range of seemingly lifeless processes, from climate and atmospheric chemistry to the shape of the landscape and even, maybe, plate tectonics.

'The signature of life has gone everywhere – it's really modified the whole planet,' says Colin Goldblatt, an earth-systems scientist at the University of Victoria in British Columbia, Canada. 'If you take it away, what changes? Well, everything.'

Just for fun, then, let's assume that the worst has happened and every living thing on the planet has died: animals, plants, the algae in the oceans, even the bacteria living kilometres down in Earth's crust. All of it, dead. What happens then?

The first thing to note, actually, is what will not happen. There will be none of the rapid decomposition that befalls dead organisms today, because that decay is caused almost entirely by bacteria and fungi. Decomposition will still happen, but very slowly as organic molecules react with oxygen. Much of the dead material will simply mummify; some will be incinerated in lightning-sparked fires.

Still, the first effects of the wipeout will start to show very quickly, with the climate getting hotter and drier, especially toward the centres of continents. That's because forests and grasslands act as massive water pumps, drawing water out of the soil and releasing it into the air. With no living plants, that pump shuts down and rainfall tails off – all within a week, says climate scientist Ken Caldeira at the Carnegie Institution for Science in Stanford, California.

Water evaporating from plant leaves also helps cool the planet, as though the trees were sweating, so a drier world will quickly warm up. 'I'm guessing it might be a couple of degrees,' says Caldeira.

In some parts of the world the effect may be a lot stronger. The Amazon basin, for example, depends heavily on moisture released from plants to drive its rainfall. Without the plants, regions such as these could rapidly heat up – by as much as 8°C, says Axel Kleidon, an earth scientist at the Max Planck Institute for Biogeochemistry in Jena, Germany.

That initial spike is just the start. As the years roll by, the world will continue to get warmer as more and more carbon dioxide creeps into the atmosphere. This happens largely because there are no longer any plankton in the ocean storing carbon in their bodies, dying and sinking down to the depths. As this 'biological carbon pump' grinds to a halt, carbon-depleted surface waters quickly come into equilibrium with the carbon-rich depths, and some of this extra carbon finds its way into the atmosphere. The net result is that in as little as twenty years, atmospheric CO_2 roughly triples – enough to raise average global temperatures by about 5°C, says James Kasting, a geoscientist at Pennsylvania State University.

Plankton will be missed in another way, too, because they release large amounts of a compound called dimethyl sulphide into the atmosphere over the oceans. These molecules act as seeds for water vapour to condense into clouds – especially the low, dense clouds that help radiate heat away from the planet's surface. Without plankton, almost immediately the clouds that form over the oceans will have bigger droplets and therefore be darker, absorbing more heat, says Caldeira. That could contribute another 2°C of warming within years to decades. On top of the 5°C from all the extra CO_2, that would be enough to rapidly accelerate the melting of the polar ice caps.

As the world warms, more water will evaporate from the oceans, so more rain will fall. Not everywhere will get wetter, though. Most of the extra rain is likely to fall where it does today – in equatorial regions where converging winds cause air to convect upward, cool and dump its moisture. Wet places are likely to get wetter and desert regions are likely to get drier – not that there will be any living things around to care.

While all this is happening, Earth will gradually be stripped of soil. No longer held in place by a mat of plant roots, it will be washed away. In hilly environments with plenty of rain, this could take centuries. Flatter landscapes might take considerably longer. In places like the Amazon basin it could take tens of thousands of years, says geomorphologist William Dietrich at the University of California, Berkeley.

All that eroded soil has to go somewhere, and most of it will end up in the ocean, in vastly larger deltas and outwash fans at the mouths of the rivers that take it there.

Rivers, too, will change. The deep-banked, meandering rivers so familiar to us today depend on plant roots to slow the erosion of their banks and keep them from spreading over the landscape. When those roots vanish, rivers will begin cutting through their banks, transforming from a single main channel into a network of braided streams like those seen today in deserts or at the foot of glaciers, says Peter Ward, a geologist at the University of Washington in Seattle. The world has seen this before: during the Permian mass extinction, about 250 million years ago, rivers abruptly changed from meandering to braided.

As the soil disappears, the world will also become sandier. The finer clay sediments so common today are largely a by-product of worms and other organisms that physically break up soil. Without these, the main mechanism for breaking up bedrock will be freeze/thaw cracking and wind erosion, so fragments will be fewer and coarser.

That seemingly small change in particle size, summed over hundreds of thousands of years, will have two big effects. The easiest to see will be changes to the landscape. Larger, coarser particles make for more abrasive sediments in streams and rivers. Over time, these make waterways cut a steeper path to the ocean, and as they steepen so too do the valley slopes. 'It's easy to imagine that you would go to a more jagged landscape,' says Peter Molnar, a geologist at the University of Colorado in Boulder.

Such stream cutting could also get a boost from changes in run-off patterns. Even though less rain and snow are likely to fall on inland regions, the lack of soil to hold the moisture may mean that any that does fall will run off as flash floods. Since most erosion happens during torrential flows, this could mean that in some places rivers will cut down into bedrock more sharply than today even though they carry less water on average, says Taylor Perron, a geomorphologist at the Massachusetts Institute of Technology.

In other places, though, less rain and snow – and fewer glaciers, the swiftest agents of valley-cutting – are likely to lead to less erosion. Whichever way the erosion balance tips, over millions of years it could change the height and shape of mountain ranges by altering the equilibrium between mountain-building and erosion. 'It's not too poetic to say that trees matter to mountains,' says Dietrich.

Still, these changes will be relatively subtle ones. As photos from the surface of Mars show clearly, a world without life wouldn't appear all that strange to us. 'You look at them and you think, well, Arizona, New Mexico,' says Dietrich. 'There'd be lots of rock and very little soil. But it wouldn't feel like a foreign planet.'

Not unless you check the thermometer, that is, because the increase in the size of eroded and sedimentary particles will make a surprisingly big difference to the climate by reducing the rate of chemical weathering of rock – a key feedback in the planet's climate control. Chemical weathering refers to a reaction between silicate rocks and CO_2 to form carbonate compounds. Eventually, these carbonates find their way to the ocean floor, where the carbon is locked away as limestone. Since living things break up bedrock into fine particles, they increase the rock's total surface area and so speed up chemical weathering.

Exactly how much is an open question, but what little evidence there is suggests that life raises weathering rates ten to a hundred-fold, says David Schwartzman, a biogeochemist at Howard University in Washington, DC. With less weathering, atmospheric CO_2 levels will rise until weathering rates equilibrate again. Over the course of a million years or so, CO_2 levels could increase enough to raise average temperatures from around 14°C today to 50°C or even 60°C, Schwartzman estimates. That is easily enough to do away with all the ice caps.

At the same time CO_2 is building up, oxygen will be disappearing. The early Earth was almost devoid of molecular oxygen, which is far too reactive to survive without steady replenishment. Only after photosynthesis began generating oxygen some 2.6 to 3 billion years ago did this gas begin to accumulate in the atmosphere. After life's demise, it would gradually ebb away. Within about ten million years, the atmosphere is likely to contain less than 1 per cent of the oxygen it does now, says planetary scientist David Catling at the University of Washington, Seattle.

At this point there will be too little oxygen to maintain the ozone layer. Without this protective blanket, Earth's surface will be blasted by ultraviolet light. 'It would start to be bad in terms of ultraviolet light after ten to twenty million years,' says Catling.

Loss of oxygen will also make the planet a drabber place. Iron-rich rocks will no longer oxidise to their familiar reddish colour. 'The surface would get greyer,' says Kasting. But there will be highlights. Shiny minerals such as pyrite and uraninite, which form in low-oxygen environments and were common on the early Earth, will resume forming.

An oxygen-free atmosphere rich in CO_2, naked bedrock exposed at the surface of the continents, minerals last formed billions of years ago – that all sounds oddly familiar to earth scientists. 'If you killed off life and waited a hundred million years, my guess is that it would look a lot like the planet would if there had never been life,' says Caldeira. Others echo his hunch.

Earth's lifeless future may differ from its lifeless past in one important way, though. The sun was about 30 per cent fainter at the beginning of Earth's existence and has been brightening ever since, so the abundant CO_2 in the atmosphere would have been a plus, helping to keep the early Earth from freezing over. Under our hotter modern sun, CO_2 is likely to push Earth into a more extreme state.

In fact, Goldblatt thinks that losing life could tip the climate balance entirely. Some models suggest that if temperatures rise enough, the increased humidity in the atmosphere could trigger a runaway greenhouse effect in which higher temperatures lead to more atmospheric water vapour – a potent greenhouse gas – which could raise temperatures still further in a vicious cycle. 'Earth today is probably reasonably near that threshold,' he says.

It doesn't look like human-made climate change could push us there, Goldblatt stresses. 'We're talking about much bigger changes. But we've got millions of years to play with, so I think it's realistic that we could get to a runaway greenhouse.' In the extreme, temperatures could rise enough to boil away the oceans, so that the planet could ultimately end up with surface temperatures over 1,000°C. 'It may well be that the answer to what Earth would look like without life is Venus,' he says.

Others are less pessimistic – if that is the right word when discussing a speculative event hundreds of millions of years in the future. Venus probably became such a hothouse because its plate tectonics stopped early in its evolution, says climate modeller Peter Cox at the University of Exeter, UK. He thinks that Earth, which is still tectonically lively, will continue to bury carbon through subduction of crustal plates, keeping large amounts of CO_2 out of the atmosphere and probably averting a runaway greenhouse effect.

Trouble could still be lurking around that distant corner, though, because subduction might slow down in the absence of life. Without life, there would be much less of the fine clay sediments that lubricate crustal movement at subduction zones. This could be enough to slow or even halt tectonic activity, says Norm Sleep, a geophysicist at Stanford University in California.

The long-term forecast for our hypothetical sterile Earth is not encouraging, it seems. Without its blanket of life, Earth may not look radically different, but it is likely to become a much more hostile place: hotter, steeper, bathed in radiation and with more severe extremes of rainfall. In the long run, it could end up totally uninhabitable.

Unless, of course, something remarkable happens. No one really knows how life originated the first time round, but it seems clear that it happened within a few hundred million years of the planet cooling enough to be habitable. The same could happen again soon after the extinction event. After all, most of the atmospheric oxygen – a poison to many prebiotic chemical reactions – will be gone, and there could be plenty of organic molecules lying about. Best of all, there will be no pre-existing life to gobble up those tentative early steps – a handicap that may well have prevented a second genesis on Earth.

In fact, a newly sterile Earth – a clean slate – could end up being the best gift a novel future life form could hope for.

What will the last days of life on Earth be like?

? Take a trip through the next 7 billion years with Andy Ridgway to find out what kind of animals will survive as mountains disappear, oceans spread and the planet fries.

We all know the ending: everybody dies. Since the sun's birth 4.6 billion years ago, its core has been getting ever denser and hotter. It is now 30 per cent more luminous than at birth and it's only going to get brighter. Ultimately, life's fate is sealed – it will be fried by the sun's intense energy. Earth will once more be a dead rock.

But let's not jump ahead quite that far. Take a few minutes to think about Earth's final chapter. What will be the last organisms to survive as the planet fries and where will they hide? What will our blue marble look like in its swansong millennia? Humanity will have vanished long before the final act, so we will never truly know – but that hasn't stopped researchers from making educated guesses about how it might unfold.

The end of life is not going to be a simple decline into nothing. There will be periods of resurgence when new, bizarre life forms are spawned. Mountains will stop growing. When that happens will determine what lives and what does not. And there's the question of what role we humans will play in Earth's future – in particular whether we're able to give ourselves a temporary stay of extinction.

It was James Lovelock, best known for the Gaia hypothesis, who first considered the effect of the sun's brightening on Earth. In a 1982 paper written with Michael Whitfield, the pair pointed to a known chemical reaction: carbon dioxide in raindrops reacts with silicate rocks, producing solid carbonates. This weathering process takes CO_2 out of the atmosphere – and the hotter the temperatures, the more it rains and the faster this happens. Whitfield and Lovelock wrote that as Earth warms, weathering should increase, eventually reducing CO_2 levels to such a degree that photosynthesis must cease. Sure, taking CO_2 out of the atmosphere would dampen the greenhouse effect and put the brakes on rising temperatures, but that's only in the short term. Over time this would be overwhelmed by the warming of the sun.

No photosynthesis means no plant life and no plant life is never good news for animals. Lovelock and Whitfield suggested the complete extinction of life on Earth could unfold in just 100 million years, a mere blink of an eye on geological timescales. While the basic idea has stuck, the current thinking is that it will actually take 600 to 900 million years for CO_2 concentrations to fall below the 10 parts per million necessary for photosynthesis.

Astrobiologist Jack O'Malley-James of Cornell University recently sketched out a possible fate for Earth's swansong biosphere. He teamed up with other astrobiologists and plant biologists, and used what we know about animal, plant and microbial energy needs, as well as other factors such as species' ability to move to new habitats or migrate, in order to build a rough sequence of extinctions over the next 4 billion years.

'It's a little like following the evolutionary tree of life in reverse,' O'Malley-James says. 'Animals get smaller and simpler.' Along the way, some species are expected to fare better than others. Migratory birds, for instance, are able to seek out cooler, higher regions as Earth warms up. And life in the sea should cope slightly better, since water takes longer to warm up than air.

In the simplest scheme, once large and small vertebrates have died on land and in the sea, only marine invertebrates would remain, with microbes to keep them company. O'Malley-James proposes that the last non-microscopic animals will be tube worms living around deep-sea hydrothermal vents.

Things will go from bad to worse 1 billion years from now. By then, average global temperatures are expected to reach 47°C. The oceans will rapidly evaporate. The additional water vapour in the atmosphere will trigger a runaway greenhouse effect. Microbial life will cling on in shrinking pockets of water. They will be snuffed out first in the tropics, then at the poles. For a time, mountaintops and underground ice caves will provide shelter from the baking heat. But life's final bolthole will be the deep subsurface, where microbes will continue to eke out a living until – under the most optimistic estimate – they finally disappear 3 billion years from now.

That's the idea in basic form. The reality is, of course, far more complicated and several factors could throw huge spanners in the works. For starters, recent studies suggest our understanding of rock weathering could be flawed, which would mess up O'Malley-James's timing estimates. 'While there is a link between warmer temperatures and increased CO_2 draw-down,' he says, 'other factors, such as relief, rock type and acidity, could be more important.' The net result is that CO_2 may not react as quickly as had been thought, which would mean plants carry on for longer than predicted, stalling the biosphere's collapse by hundreds of millions of years, or longer.

Plate tectonics could also transform the fate of the biosphere. The process is driven by geothermal heat, which comes from radioactive decay of isotopes deep within Earth.

But there is a finite amount of material, so the amount of energy released will slowly drop. When plate motion eventually grinds to a halt, mountains will stop rising and, over millions of years, erosion will level the land. 'That could happen any time between half a billion and two billion years from now,' says David Catling. The exact timing will govern life's final stages, and whether Earth becomes a water world before drying up.

The moon has a role to play in these events. It is moving away from us by 3.78 centimetres a year. Some time between 1.5 and 4.5 billion years from now, it will stop stabilising Earth's tilt. 'The poles will start tipping to the line where the equator would have been,' says Lewis Dartnell, an astrobiologist at the University of Westminster, UK. Without the moon's stabilising influence, Earth's tilt could swing erratically. 'That will have extraordinary climatic effects.' If there were still plants and animals, 'they probably wouldn't stick around for much longer', says O'Malley-James. 'The climatic conditions would be constantly changing. If you change things too rapidly, organisms can't evolve or adapt to the new conditions, and you are likely to get a lot of extinction.'

There's another possibility. While the moon is still exerting some control, Earth's axis could settle somewhere other than its current 23-degree tilt. If it becomes greater, bigger extremes between seasons could keep some regions habitable for longer, says O'Malley-James.

Shifts in Earth's axis and tectonics aside, it is the changes in temperature and CO_2 that give us some of the most intriguing possibilities. That's down to the fact that they are going to rise and fall in a jerky fashion, says Peter Ward, an astrobiologist and palaeontologist at the University of Washington in Seattle. Rock weathering happens seven to ten times faster when plants are around, because their roots break up rock and expose more of it to CO_2. 'But it will get to the point where complex plants die and so you lose roots and the weathering slows down,' says Ward. At the same time, volcanoes will continue belching CO_2, so levels will rise for a while. As the sun gets brighter, its luminosity will also become more erratic. A sudden increase in intensity would boost weathering and bring CO_2 back down.

Under these conditions, says Ward, Earth's biosphere will fluctuate. During cooler times, life will get a reprieve, and complex organisms could evolve again – organisms that may be quite different from what we're familiar with, specially adapted to low oxygen and warm temperatures. Bizarre body plans could evolve.

Ward imagines animals evolving new adaptations, such as shields to protect them from intense radiation – something like a turtle with a shell made of iron-rich minerals. 'Or you could almost imagine an animal that has a big bag of water on its back that would protect its inner organs, because water can also serve as a shield.'

And what of us? A look at the fossil record does not paint an optimistic picture. 'Mammal species only last about a million years on average,' says Catling. 'A species lifespan of ten million years is very rare.' We've been here for 200,000 years so far, so we still have hundreds of thousands of years ahead of us. But odds are we will be long gone before things get really hairy. 'I appreciate this is not a popular view,' says Catling. 'The conceit that humans are invincible on geologic timescales is widespread and far more popular.'

He cites disease, natural disaster and self-inflicted ecological collapse as possible curtain calls for our species. A rise of just 8°C would change civilisation as we know it, says Johan Rockström at the Stockholm Resilience Centre in Sweden. Sea levels roughly 60 metres higher than they are today would eradicate most urban centres. Fresh water supplies would shift towards the poles, leaving the tropics essentially uninhabitable. 'This would most likely mean a concentration of human populations in the southern and northern tips of the hemispheres,' says Rockström.

Let's be fanciful and imagine *Homo sapiens* overcame the rather stiff odds and found a way to cope with all this. In that case, we might well evolve to suit our new conditions, says Rockström.

What we currently see as the pinnacle of human evolution may turn out to be transient, says Catling: 'In the optimistic scenario that humans survive, technology will transform any descendants into a post-human species that will be barely recognisable to us. It's impossible to imagine how exactly advances in modern gene therapy or prosthetic devices will change the human species into another species. But another species will surely be the result.'

And who is to say what that species would be capable of. 'In the far future, if humans are still around, or some other intelligent species, they would presumably be doing everything they could to stave off the temperature rises,' says Dartnell. Options are pretty limited. 'The only thing that could really delay the planet becoming uninhabitable would be geoengineering on a truly massive scale – basically a planetary sunshade. But by four billion years from now, even maintaining a sunshade becomes problematic,' says Catling. There is, however, an even more radical alternative.

As the sun gets brighter, its habitable zone will sweep towards the edge of the solar system, to the point where Earth would no longer fall within it. 'So why not move us outwards to a wider and wider orbit so the planet stays within this migrating habitable zone,' says Dartnell. 'We could start sending comets or asteroids down towards Earth so they gravitationally slingshot past. If you arrange that encounter, you can transform the orbital energy of the comet into the orbital energy of the Earth and it will migrate outwards.'

But even that would have its limit – no civilisation could possibly withstand the sun's red giant phase 7.5 billion years from now. So ultimately, the ending is always the same – everybody dies. Unless we've moved somewhere entirely different, that is.

Could our planet escape a dying sun?

? Trapped in the sun's warm embrace, the inner planets are toast, but Earth might slip its leash. Joshua Sokol examines the ultimate escape plan.

Our star is not destined to explode as a supernova, hurling its planets into space. It's just not massive enough. But when it finally burns through its supply of hydrogen some six billion years from now, the great sphere of hot plasma at the centre of our solar system will grow so spectacularly bloated and bright that it will transform our cosmic neighbourhood for ever.

Like most stars, the sun is a main sequence star: in its core, nuclear fusion generates energy by converting hydrogen to helium. Once all the hydrogen there has been consumed, a layer of hydrogen around the core will ignite, and the extra heat produced will overcome the gravity that was keeping the sun from ballooning.

The result is a red giant: a swollen sun, thousands of times more luminous than it is now, whose outer layers will engulf the innermost planets. At full splendour, its radius will extend a little farther than Earth's current orbit.

And yet our little blue marble may escape. As the sun

swells, it will lose up to a third of its mass to a great outward wind of charged particles. With that will go some of its gravitational pull, allowing the comets, asteroids and planets held in its sway to migrate to wider orbits.

For the innermost planets, it's a race against time. 'Mercury, Venus and Earth effectively will each try to outrun the sun as it becomes larger,' says Dimitri Veras of the University of Warwick in Coventry, UK. Mercury and then Venus will almost certainly lose, each being engulfed in the sun's inflated atmosphere and torn apart by tidal forces.

The fate of Earth is less certain. As the planet drifts away, it will be hauled back in by tides from the sun's outer layers. 'The case is too close to call,' Veras says. Still, any life clinging on would be in trouble: the very tides tugging Earth inwards will cook its interior, giving rise to volcanic eruptions worldwide.

All the planets beyond Earth should survive, but their atmospheres will be transformed or boiled off. Our supercharged sun will even cause havoc in the asteroid belt, says Veras. When sunlight strikes asteroids they spin faster and faster, and many will centrifuge themselves into smithereens. The Oort cloud, a vast population of icy objects loosely bound at the farthest margins of the solar system, will quietly drift away into interstellar space.

There is a silver lining: the puffy old sun will be so luminous that the chilly outer regions of the solar system, including the Kuiper belt where Pluto resides, may become hospitable to life. But the opportunity will be fleeting.

After 800 million years as an inflated red giant, the sun will shrink to roughly eleven times its current size, then briefly swell again. Finally, its atmosphere will blow away to leave a glowing core: a white dwarf. The stellar embers will cool and eventually crystallise, leaving the Kuiper belt once again out in the cold.

Could Earth outlive the Milky Way?

? Our galaxy is set on a collision course with its neighbour. But Earth could survive unscathed, says MacGregor Campbell – if it lasts that long.

For a tiny smear of light drifting in a sea of darkness, the Milky Way seems stable enough, and indeed it has been around almost as long as the universe itself. But just as gravity created the galaxy we call home, so it has sealed its fate: a slow-dance death spiral with the nearby Andromeda galaxy.

Andromeda, also known as the spiral galaxy M31, is heading straight for us at about 110 kilometres per second. The good news is that, being more than 2.5 million light years away, it won't collide with the Milky Way for another 4 billion years.

Astronomers have known about Andromeda's approach for the best part of a century, but measurements of its trajectory weren't precise enough to tell whether our galaxy would get winged or truly clobbered. That debate is now settled. 'Our measurement implies that the encounter will be a head-on collision,' says Tony Sohn at Johns Hopkins University in Baltimore, Maryland, who has tracked Andromeda's motion in 3D using data from the Hubble telescope.

The collision itself will play out over 2.5 billion years. Andromeda will at first loom ever brighter in the night sky. Then, as hundreds of billions of stars, vast gas clouds and swathes of dark matter from the two galaxies swirl and smash, new star-forming regions will ignite, each lasting for millennia.

The galaxies will pass through each other a number of times as they merge into a new mega-galaxy, sometimes called Milkomeda. But stars and planets are unlikely to crash into one another, says Sohn. The average distance between stars in the Milky Way is 4 light years, which leaves plenty of space for Andromeda's stars and planets to pass through unscathed. Believe it or not, the initial collision is likely to leave our solar system alone – although near-misses could distort gravity, disrupting planetary orbits, says Sohn.

When the churning is done, Milkomeda will probably settle down as an elliptical galaxy – a giant ball of diffuse light in the night sky. The galactic merger will be complete, leaving a slightly larger smear of light in the endless dark.

What will happen at the end of the universe?

? How the grand finale of everything pans out depends on which strand of the multiverse you're in – and the enigmatic nature of dark energy, says Daniel Cossins.

In the 1990s, observations of distant supernovae indicated that the universe's expansion has been gathering pace over the past few billion years. 'Dark energy' is held responsible for that, but no one knows what it is. If it is unchanging, as most cosmologists assume, the cosmic ballooning will continue unabated and the universe will eventually become so thinly stretched that no galaxy, star or even particle is in contact with or even in sight of another. No new stars will form, and existing ones will burn out. As its temperature drops ever closer to absolute zero, this flaccid universe will go out in a cold, dark whimper: the 'big freeze'.

However, dark energy has made its presence felt only in the past few billion years, so it might be growing stronger over time. If so, the universe is bound for a more dramatic fate than a freeze. A surging dark energy would slowly tear apart galaxies and stars, and eventually space-time itself. Recent calculations indicate that the earliest this 'big rip' could happen is 2.8 billion years from now, well before our sun is due to burn out. Most cosmologists think the solar system is safe, however. A big rip, if it ever happens, is most likely to be tens of billions of years in the future.

If dark energy should for any reason weaken – perhaps even turn negative – gravity would finally prevail over its repellent phantom nemesis. The universe would crank into reverse gear, and begin shrinking again, right down to the same sort of pinprick of infinite density in which it started. The big bang universe will be bookended with a big crunch. Although that would be bad news for anything in the cosmos, it might not be bad for the cosmos itself: some models suggest it could rebound in a 'big bounce' that would create another universe, starting the cycle all over again.

There's a disturbing, and disturbingly possible, alternative endgame: the big slurp. The Higgs boson is the particle that gives other fundamental particles mass, and so is in some sense a guarantor of the universe's stability. But the Higgs boson discovered at CERN in 2012 is strangely light, suggesting that the universe it builds is an unstable 'false vacuum' state, teetering on the brink of ruin. A quantum fluctuation could at any moment conjure up a bubble of true vacuum. In that case the universe would eat itself from the inside out at the speed of light – faster than we would ever know.

Epilogue

So, with a heavy heart, we shut the door on the multiverse – for now. Our journey has taken in some incredible sights – an Earth without humans, another where Einstein and Newton were mere footnotes in history, and others where our successors have drained seas and become new kinds of human.

As we return to our own universe, we have to leave all these things behind, but we do return with a set of fresh perspectives. Touring the highs and lows of these alternate realities has given us enough distance to examine our own place in the multiverse. We know that nation-states are little more than a bureaucratic fudge, that organised religion could offer good without the god, and that if time was reversed, we probably wouldn't know it.

These are insights we can apply to our own universe, encouraging us to build more diverse communities, find common ground with people of different faiths, or simply stop worrying about the future – seeing as it may already have happened. Like any good journey, thinking about the multiverse has broadened our horizons, without us ever leaving our seats.

And if it seems humdrum coming home to your own universe, devoid of replicators or brain uploaders or Mars bases, just remember that our world is someone else's fantastical what-if scenario. Somewhere out there in the multiverse, a space-suited dinosaur is gazing down at Earth and wondering what her world would be like if small furry mammals had become an intelligent species instead.

You're already living in the universe next door.

Welcome back.

Acknowledgements

This book would not exist in any strand of the multiverse were it not for the combined efforts of a great many people: the writers whose stories we've selected, their editors, Chris Simms and the sub-editors' desk, Jeremy Webb, Graham Lawton and Sumit Paul-Choudhury, Toby Mundy, Georgina Laycock and Nick Davies at John Murray, the whole *New Scientist* family, and of course, our faithful readers.

About the Contributors

Sally Adee ('Could replicators end material scarcity?') is a reporter and editor at *New Scientist* and a contributor at the Last Word on Nothing. She could do so much more if someone would just invent a self-replicator.

Gilead Amit ('What if we could see the future?') is a features editor at *New Scientist*.

Anil Ananthaswamy ('. . . But intelligence is a dead end?', 'What if we don't need bodies?') is a *New Scientist* consultant and author of *The Edge of Physics* and *The Man Who Wasn't There*. He teaches science journalism at the National Centre for Biological Sciences in Bangalore, India, and is a guest editor at the University of California Santa Cruz's science writing program.

Colin Barras ('What if the dinosaurs hadn't been wiped out?') is a *New Scientist* consultant. He holds a PhD in palaeontology, and writes regularly on human evolution and the life sciences.

A. Bowdoin Van Riper ('Could electric motors have powered the industrial revolution?') is a historian who specialises in depictions of science and technology in popular culture, and has published several books including *Imagining Flight* and *Men Among the Mammoths*.

Catherine Brahic ('Could the climate be controlled?') is an award-winning science journalist and editor at *New Scientist*, with a focus on human evolution and our natural environment.

Michael Brooks ('Can we rewrite the laws of physics without destroying the universe?', published here for the first time, '. . . And what if we don't?', 'Will we still be here?') is a *New Scientist* consultant and the author of *13 Things That Don't Make Sense, The Secret Anarchy of Science* and *Can We Travel Through Time?* He holds a PhD in quantum physics and is a regular contributor to a variety of newspapers and magazines.

MacGregor Campbell ('What if we find ET?', 'Could Earth outlive the Milky Way?', 'What if we don't need bodies?') is a science writer and video producer based in Portland, Oregon. His work has appeared in numerous magazines and online publications.

Anne-Marie Corley ('Where will we explore?') is a writer and editor based in Texas, California, and the Pacific Northwest. A veteran of the Iraq war, she practices yoga for healing, feels nostalgia for maths and physics equations, and believes that time as we know it does not exist.

Daniel Cossins ('What if we're not the only intelligent species on Earth?', 'What will happen at the end of the universe?') is a features editor at *New Scientist.*

Kate Douglas ('. . . And could we have religion without one?') is a biology features editor for *New Scientist.*

Steve Fuller ('What if the Nazis had won World War II?') is a professor of sociology at the University of Warwick, and author of books including *Dissent over Descent* and *Humanity 2.0.*

Lisa Grossman ('How will early settlers colonise Mars?') is an award-winning physics and astronomy writer based in Cambridge, Massachusetts.

Shannon Hall ('Mapping the multiverse: how far away is your parallel self?') is a science reporter for *New Scientist*, *National Geographic*, and *Discover* among others.

Jeff Hecht ('Will artificial starlight power the world?', 'Where will we live?') is a *New Scientist* consultant and the author of books including *Understanding Lasers* and *Beam: the Race to Make the Laser*. He has a B.S. in engineering from Caltech and contributes regularly to a variety of magazines.

Bob Holmes ('What if we could start over?', 'Could we save the world by going vegetarian?', 'Will human-made life forms roam the planet?, 'What will our descendants know about us?', 'What if everything died out tomorrow?') is a long-time *New Scientist* consultant and the author of *Flavor: The Science of Our Most Neglected Sense*. He lives in Edmonton, Canada, where he likes to daydream about alternative futures.

Gerald Holton ('What if Einstein had been ignored?') is a professor of physics and the history of science at Harvard University.

Rowan Hooper ('Should you care about your parallel lives?') is a senior editor at *New Scientist*. He has a PhD in evolutionary biology and worked in Japan for 8 years as a scientist and then as a journalist, before taking a position at Trinity College Dublin. He eventually settled in a stable part of the multiverse, in London.

Joshua Howegogo ('What if we decide that God exists?') is a features editor at *New Scientist* where he covers physical science. He holds a PhD in chemistry and definitely believes in God – at least, he does in this universe.

Christopher Kemp ('What would Earth be like without us?', 'Could 100 babies left on an island rebuild civilisation?', 'What will become of the last humans?') is a writer based in Grand Rapids, Michigan.

Graham Lawton ('What if Einstein had been ignored', 'What will future humans be like?') is executive editor of *New Scientist* and the author of *The Origin of (Almost) Everything*. He has a BSc in biochemistry and a MSc in science communication.

Michael Le Page ('What would a world without fossil fuels look like?', 'Will genetically engineered people conquer the world?', 'Where will we live?') is a writer and editor at *New Scientist*.

William Lynch ('What if the Enlightenment had spluttered out?') is an associate professor at Wayne State University, Michigan.

Debora MacKenzie ('Is there an alternative to countries?', 'Are all societies doomed to collapse?') has been a reporter for *New Scientist* for more than 30 years, and has been called the doomsday correspondent. She writes about disease, weapons of mass destruction, limits to food production, and the evolution and occasional fragility of complex systems such as human societies.

Michael Marshall ('What if we could redesign the planet?', 'Will there be any nature left?') is a freelance science writer who has worked for *New Scientist* and the BBC. He has written extensively about scary environmental problems, plus a bunch of stories about weird animal sex.

Katia Moskvitch ('Could there be a doorway to the multiverse in our backyard?') is a former BBC science and technology journalist, *Nature* reporter and *New Scientist* contributor. Currently, she is the editor of *Professional Engineering* magazine. In a parallel freelance world, though, she still regularly writes about physics and astronomy, contributing to *Nautilus* and *Quanta* magazines.

Hazel Muir ('What if Earth didn't have a moon?') is a freelance science writer for magazines including *BBC Sky at Night* and *New Scientist.*

Sean O'Neill ('. . . But could do nothing about it?') is a *New Scientist* editor whose career has swerved wildly from refuse collection to TV production to copywriting to subediting – and finally editing. He currently commissions people-focused content but keeps his writing hand warm with occasional forays into features.

Fred Pearce ('What if the population bomb implodes?') is a *New Scientist* consultant. His books include *Peoplequake: Mass Migration, Ageing Nations* and *The Coming Population Crash.* As a baby-boomer, he has had a ringside seat as the world defuses the population bomb.

Andy Ridgway ('What will the last days of life on Earth be like?') is a senior lecturer in science communication at University of the West of England, Bristol, and an award-winning journalist.

David Robson ('How will our language evolve?') is a writer on brains, bodies and behaviour. He is currently writing *The Intelligence Trap* for Hodder & Stoughton.

Peter Rowlands ('Could anyone but Newton have put the heavens in order?') is a chartered physicist and the author of 10 books on science and history including *Zero to Infinity: The Foundation of Physics.*

Joshua Sokol ('What if time flows backwards in some universes?', 'What if we could see the future?', 'Could our planet escape a dying sun?') is a freelance science journalist in Boston. His work has appeared in *New Scientist, The Atlantic,* and *The Wall Street Journal.*

Henry Spencer ('What if we'd never stopped travelling to the moon?') is a Unix systems programmer at NASA, space historian, and long-time space enthusiast.

Max Tegmark ('Is consciousness just a state of matter?') is a professor of mathematics at the Massachusetts Institute of Technology and the scientific director of the Foundational Questions Institute.

John Waller ('What if Darwin had not sailed on the *Beagle*?') is associate professor of the history of medicine at Michigan State University. He has published several books, including *Einstein's Luck* and *Discovery of the Germ.*

Toby Walsh ('Will the rise of robots put humans in second place?') is professor of artificial intelligence at the University of New South Wales.

Richard Webb ('Could a theory of everything reveal the multiverse?', 'Will we run out of resources?') is chief features editor at *New Scientist.*

The Universe Next Door was edited by **Frank Swain**. In this version of the universe he is communities editor at *New Scientist* and the resident office cyborg.

Further Reading

Reality Is Not What It Seems: The Journey to Quantum Gravity by Carlo Rovelli (Penguin, 2016)

The Big Picture: On the Origins of Life, Meaning, and the Universe Itself by Sean Carroll (OneWorld Publications, 2016)

Cosmic Coincidences by John Gribbin & Martin Rees (CreateSpace Independent Publishing Platform, 2015)

The Long Earth by Terry Pratchet and Stephen Baxter (Doubleday, 2016)

Our Mathematical Universe: My Quest for the Ultimate Nature of Reality by Max Tegmark (Penguin, 2014)

What the Earth Had Two Moons? by Neil F. Comins (St. Martin's Press, 2010)

Never Pure: Historical Studies of Science as if It Was Produced by People with Bodies, Situated in Time, Space, Culture, and Society, and Struggling for Credibility and Authority by Steven Shapin (John Hopkins University Press, 2010)

Field Guide to Dinosaurs: A Time Traveller's Survival Guide by Steve Brusatte (Quercus, 2009)

Masters of the Planet: The Search for Our Human Origins by Ian Tattersall (Macmillan Science, 2012)

What If?: Serious Scientific Answers to Absurd Hypothetical Questions by Randall Munroe (John Murray, 2014)

The Impact of Discovering Life beyond Earth by Steven J. Dick (Cambridge University Press, 2015)

Aliens: Science Asks: Is There Anyone Out There? edited by Jim Al-Khalili (Profile Books, 2016)

Phi: A Voyage from the Brain to the Soul by Giulio Tononi (Pantheon, 2012)

Deviate: The Science of Seeing Differently by Beau Lotto (Weidenfeld &Nicolson, 2017)

Beyond Boundaries: The New Neuroscience of Connecting Brains with Machines - And How It Will Change by Miguel Nicolelis (St Martin's Press, 2012)

The Knowledge: How to Rebuild Our World After An Apocalypse by Lewis Dartnell (The Bodley Head, 2014)

How Many Friends Does One Person Need?: Dunbar's Number and Other Evolutionary Quirks by Robin Dunbar (Faber & Faber, 2010)

Scale: The Universal Laws of Life and Death in Organisms, Cities and Companies by Geoffrey West (Weidenfeld & Nicolson,)2017

Soonish: Ten Emerging Technologies That'll Improve and/ or Ruin Everything by Kelly and Zach Weinersmith (Particular Books, 2017)

Earth in Human Hands: Shaping Our Planet's Future by David Grinspoon (Grand Central Publishing, 2016)

The Earth After Us: What Legacy Will Humans Leave in the Rocks? by Jan Zalasiewicz (Oxford University Press, 2008)

Global Catastrophic Risks by Nick Bostrom and Milan M. Cirkovic (Oxford University Press, 2008)

Collapse: How Societies Choose to Fail or Survive by Jared Diamond (Viking Books, 2005)

Index